Advances in Intelligence and
Security Informatics

Advances in Intelligence and Security Informatics

Wenji Mao and Fei-Yue Wang

ZHEJIANG UNIVERSITY PRESS
浙江大学出版社

ELSEVIER

AMSTERDAM • BOSTON • HEIDELBERG • LONDON
NEW YORK • OXFORD • PARIS • SAN DIEGO
SAN FRANCISCO • SINGAPORE • SYDNEY • TOKYO

Academic Press is an imprint of Elsevier

Academic Press is an imprint of Elsevier
The Boulevard, Langford Lane, Kidlington, Oxford OX5 1GB, UK
225 Wyman Street, Waltham, MA 02451, USA

First edition 2012

Notice
No responsibility is assumed by the publisher for any injury and/or damage to persons or property
as a matter of products liability, negligence or otherwise, or from any use or operation of any
methods, products, instructions or ideas contained in the material herein. Because of rapid
advances in the medical sciences, in particular, independent verification of diagnoses and drug
dosages should be made.

British Library Cataloguing in Publication Data
A catalogue record for this book is available from the British Library

Library of Congress Cataloging-in-Publication Data
A catalog record for this book is available from the Library of Congress

ISBN–13: 978-0-12-37200-2

Printed and bound in the UK

12 13 14 15 16   10 9 8 7 6 5 4 3 2 1

Contents

Preface

Recent years have witnessed tremendous demand for integrated research at the intersection of information technologies and security related applications in recent years. Intelligence and security informatics (ISI) is aimed at developing advanced information technologies, systems, algorithms, and databases for national, international and homeland security related applications, through an integrated technological, organizational, and policy-based approach. The topics in ISI include data management, data and text mining for security informatics applications, terrorism informatics, deception and intent detection, terrorist and criminal social network analysis, public health and bio-security, crime analysis, cyber-infrastructure protection, transportation infrastructure security, policy studies and evaluation, information assurance, and many others.

Information technologies have progressed significantly in recent years. Meanwhile, national and international security is becoming increasingly sophisticated and diversified. To meet the challenges faced with security informatics, new directions in ISI research and applications have emerged to address complicated problems with advanced technologies. This book provides a comprehensive and interdisciplinary account of the new advances in the ISI domain along three fundamental dimensions: Methodology, Technological development and Application. In Chapter 1, we discuss research frameworks and methodological issue for ISI research, and outline a social computing based ISI research framework. In Chapters 2–5, we focus on security-related behavioral modeling, analysis and prediction, via agent modeling, computational experiment and plan-based inference. Chapters 6–8 present the synthesis of socio-cultural computing and security informatics research with case studies.

This book intends to benefit researchers as well as security professionals, intelligence experts and policy makers who are involved in cutting-edge research and applications in security informatics and related fields. The audience of this book includes: (i) researchers in security informatics, behavioral modeling, knowledge management and socio-cultural computing; (ii) public and private sector practitioners in the national, international and homeland security area and public policy analysts; (iii) graduate level students in information and computer science related major, intelligent systems, security informatics, and social and economic computing.

We hope that the perspectives, models, technological development, and empirical findings presented in this book will provide an integrated account of the new advances in intelligence and security informatics. We also hope that this book will contribute to the synthesis of the interdisciplinary research in ISI domain, and promote community-building among researchers and practitioners in this exciting field.

Wenji Mao
Institute of Automation
Chinese Academy of Sciences
Beijing, China

Fei-Yue Wang
National University of Defense Technology
Changsha, China

Acknowledgements

We would like to acknowledge support from the following National Natural Science Foundation of China grants: NNSFC Information Technology Research Grant #60875028, 'Studies on the Computational Model of Social Inference based on Psychological Attribution Theory'; NNSFC Information Technology Research Grant #61175040, 'Studies on the Modeling of Action Knowledge and Behavior Prediction Method Based on Web Media Information'; NNSFC Information Technology Research Grant #60921061, 'Intelligent Control and Computational Intelligence Methods and Applications'; and NNSFC Major Research Plan through Grant #70890084, 'Studies on Complexity and Security Issues and Development of Social Experiment Platform in Emergent Electronic Commerce'.

CHAPTER 1

Intelligence and Security Informatics

Research Frameworks

Intelligence and security informatics (ISI) is defined as the development of advanced information technologies, systems, algorithms, and databases for international, national, and homeland security-related applications, through an integrated technological, organizational, and policy-based approach [1]. Traditionally, ISI research and applications have focused on information sharing and data mining, social network analysis, infrastructure protection, and emergency responses for security informatics.

With the continuous advance of information technologies and the increasing sophistication of national and international security, in recent years new directions in ISI research and applications have emerged that address the research challenges with advanced technologies. To meet the challenges and achieve a methodology shift in ISI research and applications, we propose a social computing-based research paradigm consisting of a three-stage modeling, analysis, and control approach that researchers have used successfully to solve many natural and engineering science problems.

1.1 Research Methodology and Frameworks for ISI

In the past few years, ISI has experienced tremendous growth and made significant progress in academic research and practice involving both government agencies and industry. There have been a number of research methods and frameworks proposed to support intelligence and security informatics-related studies.

The Dark Web project developed one of the best-known ISI research frameworks. The project has for several years collected a wide variety of data relating to, and emanating from, extremist and terrorist groups. These data have included websites, multimedia material linked to the websites, forums, blogs, virtual world implementations, etc. The early system design contained three modules: data acquisition, data preparation, and system functionality. To further support its potential use as an open-source intelligence tool, several improvements in system functionality were later made [2].

New advances in ISI have emerged in recent years, especially the advancements in social-cultural computing. Subrahmanian [3] argues that cultural, economic, political, and

Advances in Intelligence and Security Informatics. DOI: 10.1016/B978-0-12-397200-2.00001-4

religious factors can help policymakers predict the behavior of radical groups and proposes a cultural reasoning framework, CARA (Cognitive Architecture for Reasoning about Adversaries) [4]. CARA is comprised of four components, namely a Semantic Web extraction engine to elicit organization data, an opinion-mining engine that captures the group's opinions, an algorithm to correlate culturally relevant variables with the actions the organization takes, and a simulation environment within which analysts and users can experiment on hypothetical situations. When a new event occurs, CARA can forecast the k most probable sets of actions the group might take in a few minutes. To further develop CARA, several algorithms have been proposed to forecast the behavior of radical groups [5,6].

Social computing refers to the computational facilitation of social studies and human social dynamics, as well as the design and use of information and communication technologies that consider the social context. As a new paradigm of computing and technology development, social computing can help us understand and analyze individual and organizational behavior and facilitate ISI research and applications in many aspects. To meet the challenges and achieve a methodology shift in ISI research and applications, we propose a social computing-based research paradigm called the ACP approach. Below, we introduce the three major components of ACP—namely, artificial societies for modeling (A), computational experiments for analysis (C), and parallel execution for control (P)—and discuss the philosophical and physical foundations of the ACP approach.

1.2 The ACP Approach

To achieve a paradigm shift in ISI research and applications, we propose adapting the three-stage modeling, analysis, and control approach that researchers have used successfully to solve many natural and engineering science problems [7,8]. In this section, we describe what we call the ACP approach [8–12]: artificial societies for modeling, computational experiments for analysis, and parallel execution for control.

1.2.1 Modeling with Artificial Societies

There are as yet no effective, widely accepted methods for modeling complex systems, especially those involving human behavior and social organizations. Agent-based artificial societies or general artificial systems may be the most promising approach.

Modeling with artificial societies has three main parts: agents, environments, and rules for interactions. In this modeling approach, the accuracy of approximation to real systems is no longer the only objective, as it is in traditional computer simulations. Instead, the model society represented by an artificial system is considered real—an alternative possible realization of the target society. Along this line of thinking, the real society is also one possible realization. So, the behaviors of two societies, real and artificial, are different but are considered equivalent for evaluation and analysis.

Of course, modeling with artificial systems does not exclude exact descriptions of target systems. Actually, approximation with high accuracy is still the desired goal for many applications when it is achievable. The idea of equivalent behaviors is a forced compromise that recognizes intrinsic limits and constraints when dealing with complex systems.

1.2.2 Analysis with Computational Experiments

Traditionally, social studies often use passive observations and statistical methods because conducting active tests and evaluations, let alone repeatable experiments, is difficult. Even when experiments are permissible, too many subjective, uncontrollable, and unobservable process factors can limit the validity and use of the corresponding results and conclusions. Because analytical reasoning can solve very few social computing problems, finding an effective way to conduct experiments becomes critical for further development of ISI research and development.

Modeling with artificial societies shows promise for this purpose. Using artificial societies, we can treat computers as social laboratories. We can design and conduct controllable experiments that are easy to manipulate and repeat; we can then evaluate and quantitatively analyze various factors in security informatics problems. These computational experiments are a natural extension of computer simulation techniques [10]. They require attention to basic design issues relating to calibration, analysis, and verification. They also follow design principles such as replication, randomization, and blocking, as do experiments in the physical world.

Researchers must address several important issues before we can use computational experiments effectively and widely in security-related problem analysis. These issues include how to use agents to sample and interview a population, what polling techniques to adopt, and how to manage temporal–spatial distributions in virtual worlds.

1.2.3 Control Through Parallel Execution

By parallel execution, we mean one or more artificial systems running in parallel with a real system. This is a generalization of the industrial controllers applied in conventional automation, which use analytical models to drive targeted physical processes to desired states. Parallel execution provides a mechanism for the control and management of complex social systems through comparison, evaluation, and interaction with artificial systems. As outlined in Figure 1.1, it involves three major modes of operation:

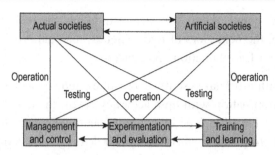

Figure 1.1 Parallel Execution for Control and Management in the ACP Approach.

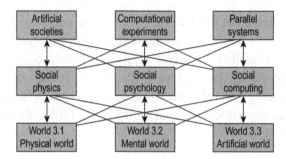

Figure 1.2 Logical and Disciplinary Foundations.

- Training and learning, where real and artificial systems are normally disconnected.
- Experimentation and evaluation, where the connections between parallel systems are on and off alternatively.
- Management and control, where artificial systems try to emulate the real system such that their behaviors can be used to improve and optimize the actual process's performance in real time.

Parallel execution can use methods and algorithms developed in simulation-based optimization and adaptive control, such as rolling-horizon analysis and model-reference feedback control. Previously, we proposed management based on parallel execution for complex social engineering systems such as transportation systems, electrical power grids, ecosystems, and social economic systems [11]. It can handle the fast dynamics and extreme scale of cyber-physical systems, making it suitable for conducting and implementing Web-related ISI research and applications.

1.2.4 Foundations in Philosophy and Physics

Figure 1.2 illustrates some ideas from philosophy and physics that could provide logical and disciplinary foundations for research in this direction.

Karl Popper, one of the greatest modern philosophers, proposed a three-world model of knowledge:

- World 1: The physical world of knowledge.
- World 2: The mental world.
- World 3: The expressed or stated world of knowledge where humankind chooses artifacts to represent knowledge, which triggers further thought.

He further divides world 3 into three parts: world 3.1 for objects in world 1, world 3.2 for those in world 2, and world 3.3 for artifacts unknown to both world 1 and world 2. Social physics, introduced by Harvard linguist George Zipf, was an early attempt to establish physical laws for world 3 as we know them from world 1 [13]. Just as natural physics provides laws governing motion in world 1, social physics will offer the foundation for constructing artificial societies. We could also view the parallel-execution concept as a practical utilization of the many-worlds interpretation that Hugh Everett envisioned for quantum mechanics 50 years ago [14].

1.3 Outline of Chapters

This monograph investigates new advances in ISI research and applications, and addresses emerging ISI research challenges with advanced technologies. It emphasizes behavioral modeling, analysis and prediction, and the synthesis of ISI and other new emerging interdisciplinary areas in three fundamental dimensions: (1) methodological issues in the security informatics domain; (2) new technological developments to support security-related modeling, analysis, prediction, and assessment; and (3) applications and integration in interdisciplinary socio-cultural fields. Therefore, we separate the contents into three groups. Following the above research framework, Chapters 2–5 discuss security-related behavioral modeling, analysis, and prediction, while Chapters 6–8 discuss socio-cultural computing and security informatics.

Open-source intelligence is becoming more and more important in security-related applications. Effective extraction of valuable information from intelligence sources is one of the central issues in the security domain. Chapter 2 presents a knowledge extraction-based computational approach to facilitate agent-based behavioral modeling via the extraction of causal knowledge and social behavior from online textual data and through the construction of causal scenarios. Chapter 3 focuses on extracting security-related event elements and generating story representation of the events, in order to facilitate computational experiments on individual and organizational behavior analysis. Case studies in the security informatics domain show the effectiveness of the proposed computational methods.

As many applications could benefit from forecasting an entity's behavior for decision making, assessment, and training, group behavior prediction has gained increasing attention in ISI research in recent years. Chapter 4 presents a probabilistic plan inference approach to forecasting group behavior. The proposed approach explicitly takes the observed agent's preferences into consideration, and computes expected plan utilities to disambiguate competing hypotheses. We use online group data to construct the domain plan library and empirically evaluate the approach in group behavior prediction. The experimental results show that our approach is compatible with human intuition in recognizing intentions and goals of the observed entities and effective in the security informatics domain. Chapter 5 further develops the multiple plan recognition method for complex group behavior forecasting.

With the advance of Internet and Web technologies, the prevalence of rich media contents, and the ensuing social, economic and cultural changes, computing technology and applications have evolved quickly over the past decade. Chapter 6 takes a synthetic view of ISI research and introduces social computing, an emerging interdisciplinary field, and its implications to research and applications in security informatics. Social computing represents a new computing paradigm. We argue that a social computing-based framework can help us develop a comprehensive ISI research framework and deal with a number of ISI research challenges.

As a case study to synthesize security informatics and social computing, Chapter 7 proposes a new social computing-based research paradigm for studying cyber-enabled social movement organizations (CeSMOs). A CeSMO is an emerging class of organizational actor playing an increasingly important role in the cyber-physical space. CeSMOs' activities can potentially have major political, security, economic, and societal implications. We discuss the defining characteristics of CeSMOs and argue that these characteristics make systematic investigation of CeSMOs in the traditional ISI research framework extremely difficult. Future research issues are also discussed in this chapter. Cultural modeling aims to develop behavioral models of groups and analyze the impact of cultural factors on group behavior using computational methods. Chapter 8 conducts an empirical study to investigate the performance characteristics of representative machine learning methods in cultural modeling.

References

1. H.S. Chen, F.-Y. Wang, D. Zeng, Intelligence and security informatics for homeland security: Informatics, communication and transportation, IEEE Transactions on Intelligent Transportation Systems 5 (4) (2004) 329–341.
2. H. Chen, D. Denning, N. Roberts, C.A. Larson, X. Yu, C. Huang, The Dark Web Forum Portal: From multi-lingual to video, Proceedings of the 2011 IEEE International Conference on Intelligence and Security Informatics, 2011, pp. 7–14.
3. V.S. Subrahmanian, Cultural modeling in real time, Science 317 (5844) (2007) 1509–1510.
4. V.S. Subrahmanian, M. Albanese, M.V. Martinez, D. Nau, D. Reforgiato, G.I. Simari, et al, CARA: A cultural-reasoning architecture, IEEE Intelligent Systems 22 (2) (2007) 12–16.

5. V. Martinez, G.I. Simari, A. Sliva, V.S. Subrahmanian, CONVEX: Similarity-based algorithms for forecasting group behavior, IEEE Intelligent Systems 23 (4) (2008) 51–57.

6. S. Khuller, V. Martinez, D. Nau, G. Simari, A. Sliva, V.S. Subrahmanian, Finding most probable worlds of probabilistic logic programs, Proceedings of the First International Conference on Scalable Uncertainty Management, 2007, pp. 45–59.

7. F.-Y. Wang, S. Tang, Artificial societies for integrated and sustainable development of metropolitan systems, IEEE Intelligent Systems 19 (4) (2004) 82–87.

8. F.-Y. Wang, A computational framework for decision analysis and support in ISI: Artificial societies, computational experiments and parallel systems, Proceedings of the 2006 International Workshop on Intelligence and Security Informatics, 2006, pp. 183–184.

9. F.-Y. Wang, Social computing: Concepts, contents and methods, International Journal of Intelligent Control and Systems 9 (2) (2004) 91–96.

10. F.-Y. Wang, Computational experiments for behavior analysis and decision evaluation in complex cystems, Journal of Systems Simulation 16 (5) (2004) 893–897.

11. F.-Y. Wang, Parallel management systems: Concepts and methods, Journal of Complex Systems and Complexity Science 3 (2) (2006) 26–32.

12. F.-Y. Wang, Toward a paradigm shift in social computing: The ACP approach, IEEE Intelligent Systems 22 (5; September/October) (2007) 65–67.

13. G. Zipf, Human Behavior and the Principle of Least Effort: An Introduction to Human Ecology, Hafner Publishing, 1949.

14. H. Everett III, 'Relative state' formulation of quantum mechanics, Reviews of Modern Physics 29 (3) (1957) 454–462.

CHAPTER 2

Agent Modeling of Terrorist Organization Behavior

Over the past decade, terrorism has become the most significant threat to national and international security. In response to the challenge, governments, law enforcement and intelligence experts are increasingly interested in the modeling of organizational behavior for counter-terrorism.

In this chapter, we present automated methods to analyze behavioral information of terrorist organizations. Our aim is to find the action set of terrorist organizations and the linkages between actions and states for the modeling and analysis of organizational behavior. Thus, we employ multiple natural language processing techniques and map the interpretation of action knowledge to concrete linguistic patterns for action knowledge extraction in the security informatics domain.

As there are huge amounts of social media data available online, we make use of the online raw textual data and choose a representative realistic group as an example for case study. Based on the online news, we present our computational approach to automatic action extraction and extraction of causal knowledge of actions and states. We also conduct an experimental study to empirically evaluate the effectiveness of our approach using online data of a real terrorist group.

2.1 Modeling Organizational Behavior

Action knowledge has been widely used in modeling and reasoning about an agent's behavior. Action knowledge is typically represented using plan representation, which includes domain actions and the states causally associated with the actions (i.e. action preconditions and effects) [1]. Action precondition is the condition that must be made true before action execution. Action effect is the state achieved after action execution. Actions can be either *primitive* (i.e. directly executable by agents) or *abstract*. In a hierarchical action representation [2], an abstract action may be decomposed in multiple ways and each decomposition is one realization of action execution. Different realizations of action execution are action alternatives.

Advances in Intelligence and Security Informatics. DOI: 10.1016/B978-0-12-397200-2.00002-6

Planning systems have been implemented and successfully applied to a number of complex domains. Planning algorithms are typically designed to automatically construct plans for agents, for example the partial order planner UCPOP [3], Graphplan [4], TLPlan [5], GP-CSP planner [6], and the HTN (hierarchical task network) planner [7]. Since action/plan knowledge is the prerequisite of various security-related applications in behavior modeling, explanation, recognition, and prediction, in this chapter we restrict our work to the terrorism informatics domain, making use of massive online data as sources of action knowledge extraction.

2.2 Action Extraction from the Web

Actions, either physical or communicative [8], imply a human agency. Action extraction focuses on extracting potentially useful actions from a huge number of Web texts. The procedure of action extraction works as follows.

2.2.1 Action Data Collection

Let C be a given corpus of Web texts and T be an organization structure, where T is a tree-like structure composed of the specific names of people associated with an organization. Collect the relevant sentence set S from C, so that for each sentence $s \in S$, s contains at least one (node) name $t \in T$.

2.2.2 Raw Action Extraction

For each relevant sentence $s \in S$, conduct syntax parsing[1] on s to generate its syntactic structure. Denote the subject, predicate, and object (if there is one) of a syntactic structure as n, v, and o respectively:

- If n contains a node name in T (including the organization name) and o does not, extract v and o according to Rule E1 and use their combination as a physical action.
- If both n and o contain a node name in T (including the organization name), extract v and o according to Rule E2 and use their combination as a communicative action.

E1: <subject> (NNP(T)) </subject> & <predicate> (VB.*?) </predicate> & <object> (?!(NNP(T)$))(NN.*?) </object>

E2: <subject> (NNP(T)) </subject> & <predicate> (VB.*?) </predicate> & <object> (NNP(T)) </object>

---

[1] We use the Stanford parser for syntax parsing: http://nlp.stanford.edu/software/lex-parser.shtml.

For predicates and objects, we mainly record verbs and nouns, and ignore other types such as adverbs, adjectives, articles, etc. For example, in the sentence '*Al-Qaeda leaders attend a rally in Gaza City*', the physical action extracted from this sentence is '*attend rally*'. In addition, we filter the extracted communicative actions using '*communication, intercommunicate*', '*communicate, pass*', '*state, say, tell*', '*accept, content*', '*disapprove, reject*', and '*intend, mean, think*' WordNet synsets.

2.2.3 Action Elimination

Not all the extracted verb–object pairs are actions, for example '*have capability*' and '*is friend*' are non-actions. *Stative verbs* [9] (e.g. *have* and *be*) are special kinds of verb that express states rather than actions. We remove from the raw action set those verb–object pairs with stative verbs.

2.2.4 Action Refinement

After eliminating non-actions, the action set may still have duplicate actions or actions expressing similar meanings. The refinement of the action set includes the following two steps, all referring to WordNet:

- Action trimming. We convert duplicate actions to their unique standard forms. For example, *plan attacks*, *planned attack*, and *planning attack* are converted to their standard form *plan attack*.
- Action combination. We combine actions that are syntactically different but express similar semantic meanings, for example *kidnap soldier* and *abduct soldier*. For two actions a_1 (verb v_1 and object o_1) and a_2 (verb v_2 and object o_2) with different syntactical forms, if v_1 and v_2 are synonyms (or the same) and o_1 and o_2 are synonyms (or the same), they can be combined into one action.

In particular, when o_1 and o_2 are locations that are instances of the same concept in WordNet, we view them as synonyms as well. For example, *attack Germany* and *attack India* can be combined into *attack country*.

2.3 Extracting Causal Knowledge from the Web

Causal knowledge is typically represented as action preconditions and effects. As the linkages between actions and their associated states are implied in domain texts, we focus on extracting action preconditions and effects from large-scale textual data on the Web. Effects are the states achieved after action execution. Thus, the linkages between actions and their effects are mainly causal relations. In contrast to the action–effect linkage, the connections between actions and their preconditions are mainly conditional.

Extracting causal relations has been studied in previous related research (e.g. Refs [10–13]). Khoo et al. [10] use manually constructed linguistic patterns to extract general causal information from medical domain texts. Girju [11] presents an inductive learning approach to the automatic detection of lexical and semantic constraints for extracting causal relations between two noun phrases. Persing and Ng [12] propose an approach to label the causes of incident reports based on text classification. The focus of our work is different from that of previous research in two aspects. First, instead of finding general causal relations between two clauses or noun phrases, our focus is to find the causal relations between actions and states—the specific causal knowledge required for inferring social causality. Second, we need to acquire richer knowledge types—not only causal relations, but also goals, reasons, and conditions associated with the actions.

Whereas the purpose of the previous research is to find general causal relations between two clauses or two noun phrases, our work focuses on finding causal relations between actions and states—the specific action knowledge. In this regard, the most relevant related work was done recently by Sil et al. [13]. They propose an SVM-based approach to build classifiers for identifying action preconditions and effects. As their work only tests a small number of actions all selected from one frame in FrameNet, and all the actions are treated as single verbs, the performance of their approach in a complex and open domain is unclear.

Because of the important role intentions play in conducting behavior, action effects are also pertinent to the goals and reasons of agents' behavior. Therefore, in extracting actions' effects, we consider both causation and reason/goal. We start from causative verbs and take other forms of causal expressions into consideration. We acquire richer knowledge types—not only causal relations, but also goals, reasons, and conditions associated with social behavior. In extracting action preconditions, we differentiate several types of precondition: necessity/need, permission/possibility, and means/tools. We classify the patterns into four categories based on their types and polarities.

Tables 2.1 and 2.2 show the linguistic patterns we design for extracting action preconditions and effects. To ensure the quality of the extracted causal knowledge, we prefer a rule-based approach that can achieve relatively high precision. On the other hand, as our work is based on open-source data, the recall rate could be compensated by the huge volume of online resources.

The expression of states is similar to that of actions, mainly including verbs and nouns and ignoring other types such as adverbs, adjectives, articles, etc. State refinement is also similar to action refinement. One difficulty of extracting causal relations is that some commonsense knowledge is seldom mentioned explicitly in online texts. We compensate for the missing effects associated with actions using commonsense knowledge of verbs by referring to the semantics of verbs in VerbNet. For example, action '*get visa*' has the effect '*have visa*', but

Table 2.1 Patterns for Extracting Action Preconditions

Precondition	Necessity/ Need	\<action (verb-ing+object/verb-ing) set\> require | demand | need \<precondition set\> \<node-name\> need \<precondition set\> to \<action (verb+object/verb) set\>
	Permission/ Possibility	\<precondition set\> allow \<node-name\> to \<action (verb+object/verb) set\> \<precondition set\> enable | create the possibility for \<node-name\> to \<action (verb+object/verb) set\>
	Means/ Tools	\<node-name\> use \<precondition set\> to \<action (verb+object/verb) set\> provide | supply | offer \<precondition set\> for \<node-name\> to \<action (verb+object/verb) set\> provide | supply | offer \<node-name\> with \<precondition\> to \<action (verb+object/verb) set\>
	Negative patterns	\<precondition set\> prevent | stop \<node-name\> from \<action (verb-ing+object/verb-ing) set\> \<precondition set\> disable | undermine \<node-name\> to \<action (verb+object/verb) set\> lack of \<precondition set\> prevent | stop \<node-name\> from \<action (verb-ing+object/verb-ing) set\> the shortage of \<precondition set\> disable | undermine \<action (verb-ing+object/verb-ing) set\> cannot \<action (verb+object/verb) set\> without | unless \<precondition set\>

this piece of knowledge is difficult to obtain online. By referring to the '*NP V NP*' frame of '*get*' in VerbNet, it has semantics '*has_possession(end(e), agent, theme)*', so '*have/possess visa*' is added automatically as the effect of '*get visa*'.

2.4 Construction of Action Hierarchy

Action hierarchy is constructed from the bottom up based on the acquired action knowledge. For two actions a_1 (composed of verb v_1 and object o_1) and a_2 (composed of verb v_2 and object o_2) with different syntactical forms and close semantic meanings, action hierarchy H is constructed in the following three cases, all referring to WordNet:

Table 2.2 Patterns for Extracting Action Effects

Effect	Causation	<action (verb-ing+object/verb-ing) set> bring about | lead to | result in | trigger | cause | produce | give rise to <effect set>
		<effect set> be caused | produced | triggered | brought about by <action (verb-ing+object/verb-ing) set>
		<node-name> <action (verb+object/verb) set> to cause | bring about | produce | trigger <effect set>
		<node-name> <action (verb+object/verb) set>, causing | producing | triggering | resulting in | leading to <effect set>
		<node-name> <action (verb+object/verb) set>, which | that bring about | lead to | result in | trigger | cause | produce <effect set>
		<effect set> caused | produced by <action (verb-ing+object/verb-ing) set>
		What bring about | lead to | result in | trigger | cause | produce | give rise to <effect set> be <action (verb-ing+object/verb-ing) set>
	Reason/Goal	the reason of | reason for | cause of <action (verb-ing+object/verb-ing) set > be <effect set>
		<node-name> <action (verb+object/verb) set> because of | on account of | due to <effect set>
		<action (verb-ing+object/verb-ing) set> be due to <effect set>
		<node-name> <action (verb+object/verb) set> for the purpose of | in an attempt to | in an effort to | in order to | so as to | in the cause of <effect set>

- If v_1 and v_2 have the same hypernym v and o_1 and o_2 are the same (or synonyms), create an or-node a (v and o_1) in action hierarchy H with a_1 and a_2 as its alternative sub-actions and a' (v and o_2) as its synonym.
- If v_1 and v_2 are the same (or synonyms) and o_1 and o_2 have the same hypernym o, create an or-node a (v_1 and o) in action hierarchy H with a_1 and a_2 as its alternative sub-actions and a' (v_2 and o) as its synonym.
- If v_1 and v_2 have the same hypernym v and o_1 and o_2 have the same hypernym o, it takes the following two steps to construct action hierarchy H:

- Create an or-node a_1' (v_1 and o) with a_1 as its sub-action and an or-node a_2' (v_2 and o) with a_2 as its sub-action in H, and then create an or-node a (v and o) in H with a_1' and a_2' as its alternative sub-actions.
- Create an or-node a_1'' (v and o_1) with a_1 as its sub-action and an or-node a_2'' (v and o_2) with a_2 as its sub-action in H, and then create an or-node a (v and o) in H with a_1'' and a_2'' as its alternative sub-actions.

For example, actions '*buy car*' and '*buy truck*' have the same verb '*buy*', and '*car*' and '*truck*' have the same hypernym '*vehicle*', thus action hierarchy with '*buy vehicle*' as the abstract action and '*buy car*' and '*buy truck*' as alternative sub-actions is automatically constructed. Another action hierarchy with '*get vehicle*' as the abstract action and '*buy vehicle*' and '*rent vehicle*' as its alternatives is constructed similarly. In constructing action hierarchy, the system starts from actions whose verbs and objects are in the lowest synsets in WordNet. With the process of action hierarchy construction, the system also moves up in the WordNet hierarchy.

2.5 Designing Causal Scenarios

The extracted action knowledge can be utilized to construct causal scenarios for experiments on terrorist group behavior. Here we provide several examples generated by our system to illustrate the construction of causal scenarios. The causal knowledge and actions used here for constructing scenarios are extracted from online resources, as explained in Section 2.4. Take the action 'launch attack' and its effects as an example. Two agents, a senior member and a leader of the organization, are involved in the scenarios. We design different ways to construct causal scenarios for computational experiment. Scenario 1 shows the intentions of both agents, so there is no evidence of coercion. Scenario 2 shows that the senior member is coerced to launch the attack, causing the pipeline damage and injury to people. Scenario 3 manipulates the senior member's freedom of choice. The senior member is allowed to choose from different options. We can further extend these scenarios to include indirect cases. In addition, as coercion may occur at more than one level of action hierarchy, Scenario 4 demonstrates a chain of command by adding an executor—the subordinate of the senior member.

Scenario 1 (Intention):
   E1   The leader of Alpha Organization plots to launch attack with the senior member of the organization.
   E2   The senior member says, 'Launching attack will hit oil pipeline to country W,
   E3   and it will also injure people.'
   E4   The leader answers, 'OK, launch attack right away!'
   E5   The senior member launches attack.

E6   Launching attack results in hitting oil pipeline and injuring people.

Scenario 2 (Subordination):

E1   The leader of Alpha Organization tasks the senior member of the organization to launch attack.

E2–E6 Same as in Scenario 1.

Scenario 3 (Freedom of Choice):

E1   Same as in Scenario 2.

E2   The senior member says, 'There are two ways to launch attack, a simple way and a complex way.

E3   Both will hit oil pipeline to country W, and the complex way will also injure people.'

E4   The leader answers, 'OK, launch attack either way!'

E5   The senior member chooses the complex way to launch attack.

E6   Same as in Scenario 2.

Scenario 4 (Chain of Command):

E1–E4 Same as in Scenario 2.

E5   The senior member commands the executor to launch attack.

E6   The executor launches attack.

E7   Same as E6 in Scenario 2.

2.6 Case Study on Terrorist Organization

We chose Al-Qaeda as a representative realistic group for our study. As a great number of reports about this group and its historical events are available online, we employ computational methods to automatically extract group actions and causal knowledge from relevant open-source textual data. The textual data we use is the news about Al-Qaeda reported in *The Times*, *BBC*, *USA Today*, *The New York Times*, and *The Guardian*, with a total of 953,663 sentences. Based on the textual data in 25,103 Web pages, we extract the members of the group and compute their associations based on the co-occurrence. Figure 2.1 illustrates the interconnections of the group members. The more active the members are, the bigger the corresponding nodes. The size of the node also reflects the relative importance of the member to the organization, according to the group's description (see *Wikipedia*). The red, blue, black, and green nodes correspond to the group, the leader, the senior members, and other peripheral members respectively.

Based on Figure 2.1, we specify the three-layer organizational structure of the group. The first layer denotes the leader (only 1 node), the second layer for the cadres (total 17 nodes) and the third layer for all the other members (total 221 nodes). The names of the group members in each layer are then used to extract actions and causal knowledge. The

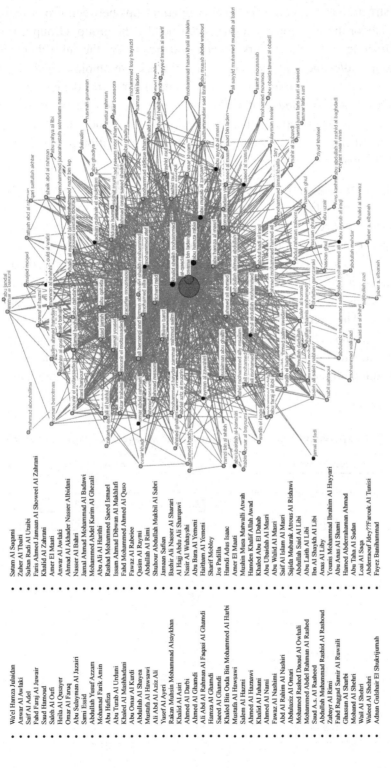

Figure 2.1 Group Members and Their Connections Extracted from Online Data.

organizational structure also specifies the power relationships between group members. In this domain, behavior includes those actions performed by the group. Group actions are acquired by extracting verb–object pairs in each sentence where the subject contains the name of the group and/or those of its associated members. We employ the designed linguistic patterns and syntax parsing (using the Stanford parser for dependencies representation) to extract knowledge of action preconditions and effects. After the refinement of the extracted group actions, action preconditions and effects, we finally collect 219 group actions (including 26 communicative actions), 146 action preconditions, and 161 action effects with quality.

Among the official investigation reports, 13 real attacks perpetrated by Al-Qaeda have relatively complete descriptions. An intelligence analyst helped us manually compose the actions and causal knowledge of each attack based on these descriptions, and these form the basis of our experiment. As our purpose is to facilitate the construction of causal scenarios and support social causality studies, we evaluate the performance of our method by checking how many actions and states specified in each attack example are already covered by the domain actions and causal knowledge we extract. Table 2.3 shows the results of the experimental study. The average coverage rates of the actions, preconditions, effects, and states (preconditions plus effects) are 85.8%, 74.1%, 78.7%, and 75.6% respectively. The performance of extracting causal knowledge is slightly lower, but given the difficulty of knowledge extraction, the results are reasonably good. In general, the experimental results

Table 2.3 Experimental Results of Causal Knowledge and Action Extraction

Attack Example	Action Converge	Precondition Converge	Effect Converge	State Converge
1	0.833	0.818	0.833	0.824
2	0.800	0.778	0.800	0.786
3	0.900	0.727	0.900	0.781
4	0.889	0.737	0.778	0.750
5	0.857	0.733	0.714	0.727
6	0.833	0.727	0.833	0.765
7	0.875	0.765	0.750	0.760
8	0.833	0.667	0.833	0.722
9	0.889	0.684	0.889	0.750
10	0.900	0.800	0.800	0.800
11	0.857	0.750	0.714	0.739
12	0.857	0.692	0.714	0.700
13	0.833	0.750	0.667	0.722
Average	**0.858**	**0.741**	**0.787**	**0.756**

verify the effectiveness of our approach in extracting causal knowledge and social behavior from Web media data, to support the construction of causal scenarios for studying social causality.

2.7 Conclusion

In this chapter, we present our computational approach to extracting causal knowledge and social behavior from online data, and conduct a case study on a typical terrorist group to evaluate the performance of our knowledge extraction method. The results verify the effectiveness of our method in extracting causal knowledge and social behavior from the Web data. Although we chose the online data of one representative group for our study, our computational method is designed to apply generally. We believe our approach is applicable to a wide range of fields. However, while we take a rule-based approach in pattern design to ensure the quality of the extracted knowledge, it is less flexible compared to statistical learning methods. Our future research will further explore the combination of automatic methods in the modeling of organizational behavior.

References

1. R.E. Fikes, N.J. Nilsson, STRIPS: A new approach to the application of theorem proving to problem solving, Artificial Intelligence 2 (1971) 189–208.
2. K. Erol, J. Hendler, D.S. Nau, UMCP: A sound and complete procedure for hierarchical task-network planning, Proceedings of the Second International Conference on Artificial Intelligence Planning Systems, 1994, pp. 249–254.
3. J.S. Penberthy, D.S. Weld, UCPOP: A sound, complete, partial order planner for ADL, Proceedings of the Third International Conference on Knowledge Representation and Reasoning, 1992, pp. 103–114.
4. A. Blum, M. Furst, Fast planning through planning graph analysis, Artificial Intelligence 90 (1997) 281–300.
5. F. Bacchus, F. Kabanza. Using temporal logics to express search control knowledge for planning, Artificial Intelligence 116 (2000) 123–191.
6. M.B. Do, S. Kambhampati, Solving planning-graph by compiling it into CSP, Proceedings of the Fifth International Conference on AI Planning and Scheduling, 2000, pp. 82–91.
7. D.S. Nau, Y. Cao, A. Lotem, H. Muñoz-Avila, SHOP: Simple hierarchical ordered planner, Proceedings of the Sixteenth International Joint Conference on Artificial Intelligence, 1999, pp. 968–973.
8. J. Austin, How to Do Things with Words, Harvard University Press, 1962.
9. D. Crystal, A Dictionary of Linguistics and Phonetics, Blackwell Publishing, 2003.
10. C. Khoo, S. Chan, Y. Niu, Extracting causal knowledge from a medical database using graphical patterns, Proceedings of the 38th Annual Meeting of Association for Computational Linguistics, 2000.
11. R. Girju, Automatic detection of causal relations for question answering, Proceedings of the ACL 2003 Workshop on Multilingual Summarization and Question Answering, 2003.
12. I. Persing, V. Ng, Semi-supervised cause identification from aviation safety reports, Proceedings of the 47th Annual Meeting on Association for Computational Linguistics, 2009.
13. A. Sil, F. Huang, A. Yates, Extracting action and event semantics from Web text, AAAI 2010 Fall Symposium on Commonsense Knowledge, 2010.

CHAPTER 3

Security Story Generation for Computational Experiments

With the advance of information technology, open-source intelligence is becoming more and more important in many applications, especially in security-related applications [1,2]. How to effectively extract valuable information from these intelligence sources so as to facilitate the generation of security stories and experimentation on security-related events has become one of the central issues in the security domain. In the last decade, there have been an increasing number of efforts on automatic knowledge discovery from massive amounts of textual data to help understand security events.

Narrative is the semiotic representation of a series of events meaningfully connected in a temporal and causal way [3]. Studies show that we organize our experience and our memory of human events mainly in the form of narrative [4]. That is, by constructing narrative about reality one can reconstruct reality [4]. Thus, generating narrative about an event (or a series of events) not only helps people understand what is going on, but also supports the experimentation on security-related events. So we choose narrative structure as a form of knowledge organization, as well as a facilitator for the computational experiments on the focused event from a considerable amount of open-source information.

As events play a central role in the security-related domain, the key problem of narrative generation is the extraction of events, as well as linking these events in a temporal way to form a narrative structure. In this chapter, we present an automatic approach to domain event extraction and story generation in the security domain. We implement a story extraction platform (SEP) that applies pattern matching to extract events from Web news and organizes events based on a theme using narrative structure. We also design extraction rules, use domain-specific features and employ ontology to facilitate story extraction in the SEP. The experimental results show the effectiveness of our system in security informatics.

3.1 Story Generation Systems

Using narrative to understand text was first proposed by Schank and Abelson [5], together with the concept of 'script'. Based on a script representation, DeJong [6] built the first story extraction system. JASPER, developed by Anderson et al. [7], is also among the early representative story extraction systems. Recently, research on narrative has been done in areas such as the description of organizational behavior [8], human–computer interaction [9,10], and automatic story generation for entertainment [11–14].

The early research on narrative is mainly focused on using narrative to understand text, but the systems developed can only extract separate pieces of information that are not well organized. Recent work shows the advantage of using narrative for different applications [9–14]. However, these do not consider the extraction of events. In the security domain, we need to combine narrative structure with event extraction in order to help both the extraction of intelligence information and computational experiments on them.

Event extraction is a traditional task in information extraction. Much work has been carried out previously. In recent years, the Message Understand Conference (MUC) [15] and Automatic Contend Extract (ACE) [16] have attracted much attention in event extraction. In MUC, the scenario template is similar to event extraction. It represents the highest level of information extraction. The performance depends on solving the core problems of the natural language process, so the best scenario template score of the last MUC is $F < 51\%$ [17,18].

Event detection and recognition have been defined as a fundamental task in the ACE evaluation plan. In ACE 2005 [19], the best score for Chinese event detection and recognition is 10.2, in contrast to that of entity detection and recognition, which is 69.2. The scores are determined by ACE and reflect the level of difficulty of each task. For the task of ACE 2005, Ahn [20–22] developed a modular system for event detection and recognition. His system works in four stages: anchor identification, argument identification, argument assignment, and event co-reference. Each stage is handled by a machine-learned classifier.

Chinese event extraction has been studied in recent years. Based on the concept of the scenario template, Yang [23] studied information extraction in the emergency domain. Wu et al. [24] developed an event extraction system based on analyzing time–space information. Similar to Ahn's work, Zhao et al. [25,26] constructed an event extraction system using trigger words for improving feature selection. Other recent work has focused on the selection and weighting of features [27–29].

Unfortunately, none of the event extraction work on Chinese texts has paid attention to the organization of the extracted event information to help understand security events and support computational experiments on them. Events are not independent of each other. In the security domain in particular, events have interconnected levels and relations. For example, given a series of events extracted from documents about one emergency, there is a theme event that represents the emergency, and a series of sub-events represent specific actions. After being organized based on their relations, events can represent the process of a security emergency. Therefore, we introduce narrative structure to organize events in the security domain.

In addition, to achieve good performance of story extraction and generation in the security domain, we employ domain-specific features in our work. We implement event detection, event element extraction, normalization and event relation extraction, and construct the story generation system and platform in the security informatics domain.

3.2 System Workflow and Narrative Structure

This section briefly describes the structure of the SEP, which is depicted in Figure 3.1.

The SEP is a modular system. It can be split into three high-level functional modules: text processing, event extraction, and narrative generation. First, in the text processing, the input is a set of raw texts about one theme event. Before event extraction, these raw texts must be preprocessed. Besides traditional word splitting and parsing, we realize a domain text-processing module for recognizing domain entities and noun phrases.

Next, the preprocessed texts serve as input for the event extraction module, which consists of four stages:

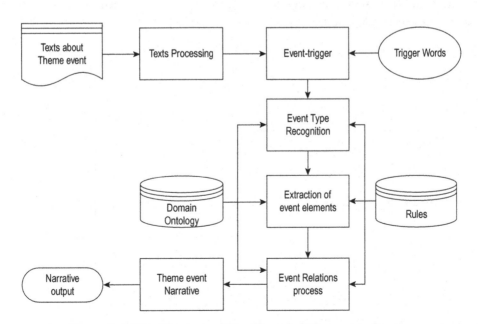

Figure 3.1 The Structure of the Narrative Generation System.

- The event trigger is used to find possible events in input texts.
- Event type recognition can detect a true event and assign an event type to it.
- Event element extraction acquires the elements of an event, such as 'when', 'where', 'who', etc.
- The event relation process is used to determine the relations between events.

Finally, all of the events will be organized according to the narrative structure. A narrative about the theme event will be generated as the output, together with the event extraction results organized within the narrative structure.

As introduced in the above section, when a security emergency occurs, there is a theme that is called the theme-event, such as the 2008 Mumbai attacks or September 11 attacks. There are also a series of specific events that we call sub-events, such as terrorists killing a man. A theme-event is a set of a series of time–space–coherence events. A sub-event is one specific event from the series. In other words, a theme-event represents the security emergency, and sub-events are those specific events that compose this theme-event. Here we define the structure of a security narrative as in Figure 3.2.

The narrative of a theme-event is a series of events arranged in the order in which they happen. The head is the time when the theme-event begins; the tail is the end of the theme-event. At each time point, there is a set of sub-events that happen at the same time. Every sub-event is composed of event elements. Type, time, and location are basic elements and are included in every type of event. Besides basic elements, different event types contain other specific elements. With this story structure, the start, development, plot, and end of the theme-event are clearly depicted.

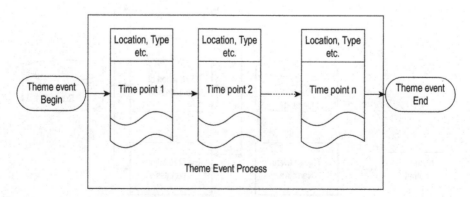

Figure 3.2 Narrative Structure.

3.3 Story Extraction Approach

In order to construct an SEP story extraction system in the security domain, we need to solve several problems. The first of these is domain text processing. While the development of general tools that can perfectly process arbitrary texts remains a future issue, currently it is possible to construct our tool to deal with domain texts with high accuracy. The second problem is event extraction. The SEP needs to extract events from open-source text, especially texts from the Web. As the annotation of massive corpora requires tremendous human labor, in such a situation we construct an event extraction module based on pattern matching and design corresponding extraction rules. The final problem is the standardization of event elements. Based on the security narrative structure, narrative generation needs type, time, and location to judge the relationship between events, so we construct domain ontology to convert time and location to the uniform form. Below we describe our approach to story extraction in detail.

3.3.1 Text Processing with Domain Knowledge

Chinese is not like English, which has a space between words. Word splitting is the basic task of Chinese text processing. In the security domain, there are many long words that are translated from other languages, such as foreign persons, locations, and organizations. In open-source text, the expression of these entities is quite free and irregular. All of these are difficult to process by a general word-splitting tool. For example, a foreign name '贾特拉帕蒂·希瓦吉' (Chhatrapati Shivaji) is too long to be recognized. To solve problems like this, we build a domain word dictionary based on the names of organizations and people in the security database of our lab.

As many long sentences are used to clearly describe the details of a security emergency, this results in unsatisfactory performance of general Chinese parsers at present. If just one word in a sentence is assigned an incorrect POS, the result for the whole sentence cannot be used. To achieve good parsing results in practice, we develop a domain text-processing module. It needs to recognize domain words, entities, and base noun phrases. The domain text-processing module is based on the suffix-driven method. Once a suffix is found, the module checks the words on the left. If they match a pattern, the results of word splitting will be adjusted. For example, the pattern '1#<0:.*?:nh:E><2:.*?:q:><3:.*?:m:B> -> <0:0:nhs>' can recognize a noun phrase whose POS is human with number and unit on the left. The left of '->' is the pattern matching part and the right is the result part. There are orders between patterns too.

Our domain text-processing module significantly improves the performance of the general text parser in the security domain. Figure 3.3 shows an example using both the general Chinese parsing [30] and our domain adjustment. '孟买警察和反恐部队逮捕一名恐怖分子

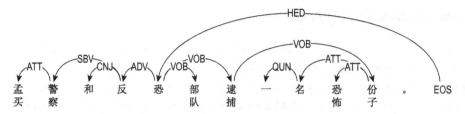

Figure 3.3 Parsing Result Using General Chinese Parser.

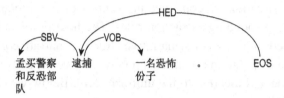

Figure 3.4 Parsing Result Using Domain Adjustment.

Table 3.1 Event Types

Type	Subtype
Dev-event	Attack, shooting, explosion, confrontation, kidnapping, person-trapped, rescue, arrest, assistance, eliminate-terrorist, event-continue, event-end
Life-event	Death, injured
Loss-event	Artifact loss, pecuniary loss
Fact	Mastermind, purpose

警察' (the police) and '反恐部队' (anti-terrorism unit) are good people in the security domain. '恐怖分子' (terrorist) is a bad person. After word splitting and domain adjustment, the sentence becomes '孟买警察和反恐部队' (the police and anti-terrorism unit of Mumbai) '逮捕' (arrest) '一名恐怖分子'(a terrorist). Their POSs are 'nhz' (noun of a good person), 'v' (verb), and 'nhf' (noun of a bad person), because the domain entity '反恐部队' is not recognized by word splitting. For example, in Figure 3.3, the general parser gives an incorrect result. Figure 3.4 shows that the result of our domain adjustment is correct and concise.

3.3.2 Event Detection and Event Element Extraction

We analyze the characteristics of security intelligence and find 18 types of events that are important and necessary in understanding emergency processes. These event types are classified into four main groups (Table 3.1):

- Dev-event is the event type that is composed of the development of the theme-event. It contains 12 subtypes.
- Life-event in the security domain contains the dead and the injured.
- Loss-event includes losses such as the damage of infrastructure or an amount of money.
- Fact is a description of the theme-event's elements. For example, a terrorist organization has claimed responsibility for an attack.

Based on the narrative structure, the most important event elements are type, time, and location. Types of events are also important in evaluating the relations between events. For example, a sub-event belongs to dev-event. Life-event and loss-event are always the result of other events.

In the SEP system, we collect and use trigger words to detect events. If a sentence contains a trigger word, an event is triggered. Many trigger words are verbs, or other special words like '死者' (dead) or '伤者' (injured). Once an event is triggered, the next step is to judge whether the event really exists in this sentence. If the sentence matches a pattern, the event is identified and a type is assigned to it. Then other elements of the event are extracted accordingly.

3.3.3 Design and Organization of Patterns

All steps of event extraction are based on pattern matching. In order to improve the coverage of patterns and use the text feature as much as possible, we introduce multilevel patterns to identify event types and elements. They are described below.

Patterns to identify event types are designed using trigger words as the core and contexts of trigger words to determine types. All other patterns have core words, and use the lexical or syntactic relations to extract event elements. Based on the different relations being used, there are two types of patterns. One type uses the context of core words and the other uses the relations in a syntax tree. All use the word and POS.

Patterns are organized with trigger words as core. All patterns that are associated with a special trigger word and contexts of it are composed of a set with the event as a unit. Every set contains type, time, and location patterns. Patterns that extract the event type are the kernel of the set and have the highest priority. There are additional patterns according to event types. For example, an injury event contains patterns that extract who are injured and their total number. There are patterns to extract sponsors and victims in a shooting event. The example below shows a set of patterns with the trigger word '丧生' (killed) as core.

■ <Event>
  <Type>死亡</Type>
  <Kernel><1><丧生:v><Rule#{L<2:.*?:[∧h]*?:><1:有:v:>E}End></Kernel>
  <Time><Rule#1 1 {到:p}{R<2:.*?:nt.*?:a>E}End></Time>
  <Location></Location>
  <Other><Rule#0 0 {R{0:<SBV><.*?><nh.*?>:naj}E}-><死亡人数及角色>[naj]End>
  <Rule#1 1 {包括:v}{R<2:.*?:nh.*?:naj>E}-><死亡人数及角色>[naj]End></Other>
  </Event>

■

Moreover, pattern matching uses flexible strategies. Each pattern can be in a positive or negative direction. The lexical type can adjust the matching distance, for example the first word on the left, the second or just the left-hand word within one clause. In summary, our pattern design and organization are efficient, concise, and have good coverage.

3.3.4 Event Element Standardization

After event extraction, the extracted elements must be normalized before they are used to generate narrative. In particular, time and location information is important in evaluating the relations between events. In the SEP, ontology is employed to solve this problem.

Number Conversion

We change from Chinese numbers to Arabic numbers. For example, '一百零五万零五百零五点五五' (one million, fifty thousand, five hundred and five point five five) is converted to 1,050,505.55.

Time Standardization

We use time ontology combined with number conversion. Time and data are converted to the standard form. When processing a word like '昨天' (yesterday), we use the publishing time of the news as the standard time. For an event without information on the time that it happened, we use the publishing time instead.

Location Standardization

Location ontology can be used to solve aliases, short names, and ambiguity. For example, '旁遮普' (Punjab) is a province in India and also in Pakistan. When Punjab is extracted, we use location ontology to compute the distances between Punjab and other extracted locations, and then choose the meaning with the shorter distance.

3.3.5 Evaluation of Event Relations

After events are extracted, narrative generation needs to evaluate the relationship between events. We use heuristic rules in our judgments:

- All of the events that are mentioned by the corpora about one theme-event are relevant to the theme-event.
- Events with time or location within the scope of the theme-event are relevant to the theme-event.
- In one sentence, if there is a dev-event, all the life-events and loss-events are regarded as the results of this dev-event. Otherwise, they are regarded as the results of the theme-event.
- The left events are the sub-event of the theme-event; these events also compose the narrative of the theme-event.

3.4 Experiment

We used Google News [31] to collect different Chinese reports on the 2008 Mumbai attacks. Based on the SEP, there are 199 patterns for domain entity recognition and 122 groups of event patterns. The experimental results of event and element extraction are given in Table 3.2.

Below, we show part of the narrative generated by the SEP:

■ Standard time: Fri Dec 26 21:09:09 CST 2008
Type: 袭击
Time: 26日 21时 过后
Location: 利奥波德餐厅
Description: 事件发起者及属性[枪手]

Standard time: Fri Dec 26 21:30:07 CST 2008
Type: 袭击
Time: 26日 晚 当地 时间 9时 30分 左右
Location: 孟买 街头 孟买 贾特拉帕蒂·希瓦吉火车站、孟买 市政府
Description: 事件发起者及属性[多 批 携带 机关枪 和 手榴弹 的 恐怖 分子]

Standard time: Fri Dec 26 22:02:05 CST 2008
Type: 枪击
Time: 26日 晚间 10点 左右
Location: ... ...

Table 3.2 Experimental Results

	Type					
	AR	*AE*	*RE*	*P*	*R*	*F*
Event	543	480	438	0.912	0.807	0.857
Element	751	588	563	0.957	0.750	0.841

AR = all possible right results, *AE* = all extraction results, *RE* = all right extraction results, *P* = precise, *R* = recall, $F = 2 \times P \times R/(P + R)$.

After time extraction and standardization, the standard time is produced and used to construct the narrative. Type, time, location, and description are the elements of events, which are all extracted from Web texts.

3.5 Conclusion

Open-source intelligence is becoming more and more important in security-related applications. In this chapter, we support computational experiments by constructing domain stories automatically from Chinese open-source texts. We implement the story extraction platform (SEP) in the security informatics domain, which applies pattern matching to extract events from Web news and organizes events based on themes using narrative structure. We also design patterns, use domain-specific features, and employ ontology to facilitate story extraction in the SEP. The experimental results show the effectiveness of the SEP system in security informatics.

References

1. F.-Y. Wang, K.M. Carley, D. Zeng, W. Mao, Social computing: From social informatics to social intelligence, IEEE Intelligent Systems 22 (2) (2007) 79–83.
2. Y. Yao, F.-Y. Wang, J. Wang, D. Zeng, Rule + exception strategies for security information analysis, IEEE Intelligent Systems 20 (5) (2005) 52–57.
3. S. Onega, J. Landa, Narratology: An Introduction. Longman, 1996.
4. J. Bruner, The narrative construction of reality, Critical Inquiry 18 (1991) 1–21.
5. R. Schank, R. Abelson, Scripts, Plans, Goals and Understanding: An Inquiry into Human Knowledge Structures, Lawrence Erlbaum Associates, 1977.
6. G. DeJong, An overview of the FRUMP system, Strategies for Natural Language Processing 113 (1982) 149–176.
7. P.M. Anderson, P.J. Hayes, A.K. Huettner, L.M. Schmandt, I.B. Nirenburg, S.P. Weinstein, Automatic extraction of facts from press releases to generate news stories, Proceedings of the Third Conference on Applied Natural Language Processing, 1992, pp. 170–177.
8. M. Fayzullin, V.S. Subrahmanian, M. Albanese, C. Cesarano, A. Picariello, Story creation from heterogeneous data sources, Multimedia Tools and Applications 33 (2007) 351–377.

9. H. Jenkins, Game design as narrative architecture, in: N. Wardrip-Fruin, P. Harrigan (Eds.), First Person: New Media as Story, Performance and Game, MIT Press, Cambridge, MA, 2004.

10. M. Mateas, Interactive drama, art and artificial intelligence, Ph.D. dissertation, Carnegie Mellon University, 2002.

11. M. Riedl, R.M. Young, From linear story generation to branching story graphs, IEEE Computer Graphics and Applications 26 (2006) 23–31.

12. M. Riedl, R.M. Young, An intent-driven planner for multi-agent story generation, Proceedings of the Third International Joint Conference on Autonomous Agents and Multiagent Systems, 2004, pp. 186–193.

13. M. Riedl, R.M. Young, Narrative generation: Balancing plot and character, Ph.D. dissertation, North Carolina State University, 2004.

14. R.M. Young, Story and discourse: A bipartite model of narrative generation in virtual worlds, Interaction Studies 8 (2007) 177–208.

15. N. Chinchor, Overview of MUC-7/MET-2, Proceedings of the Seventh Message Understanding Conference, 1998.

16. The ACE 2005 Evaluation Plan, <http://www.ldc.upenn.edu/Projects/ACE/Annotation>.

17. B. Li, Y. Chen, S. Yu, Research on information extraction: A survey, Computer Engineering and Applications 39 (2003) 1–5.

18. C. Aone, L. Halverson, T. Hampton, M. Ramos-Santacruz, SRA: Description of the IE^2 system used for MUC-7, Proceedings of MUC, 1998.

19. Q. Zhao, J. Liu, H. Feng, An ACE view of the development tendency of information extraction technology, New Technology of Library and Information Service 3 (2008) 18–23.

20. D. Ahn, The stages of event extraction, Proceedings of the Workshop on Annotations and Reasoning about Time and Events, 2006, pp. 1–8.

21. H. Ji, R. Grishman, Refining event extraction through unsupervised cross-document inference, Proceedings of the 46th Annual Meeting on Association for Computational Linguistics, 2008.

22. A.K. Irvine, Natural language processing and temporal information extraction in emergency department triage notes, Masters thesis, University of North Carolina at Chapel Hill, 2008.

23. E. Yang, On the information extraction of sudden events, Ph.D. dissertation, Beijing Language and Culture University, 2005.

24. P. Wu, Q. Chen, L. Ma, Research on extraction and integration of developing events based on analysis of space–time information, Journal of Chinese Information Processing 20 (2006) 21–28.

25. Y. Zhao, X. Wang, B. Qin, W. Che, T. Liu, Automatic event type extraction in Chinese event extraction, Third SWCL Conference, 2006.

26. Y. Zhao, B. Qin, W. Che, T. Liu, Research on Chinese event extraction, Journal of Chinese Information Processing 22 (2008) 3–8.

27. Z. Chen, H. Ji, Language specific issue and feature exploration in Chinese event extraction, Proceedings of NAACL HLT, 2009, pp. 209–212.

28. J. Fu, Z. Liu, Z. Zhong, J. Shan, Chinese event extraction based on feature weighting, Information Technology Journal 9 (1) (2010) 184–187.

29. H. Tan, T. Zhao, J. Zheng, Identification of Chinese events and their argument roles, Proceedings of the 2008 IEEE Eighth International Conference on Computer and Information Technology Workshops, 2008, pp. 14–19.

30. HIT Information Retrieval Laboratory, Language Technology Platform, <http//ir.hit.edu.cn/demo/ltp>.

31. Google News, <http://news.google.com>.

CHAPTER 4

Forecasting Group Behavior via Probabilistic Plan Inference

Group behavior prediction is an emerging research and application field in intelligence and security informatics, which studies computational methods for the automated prediction of what a group might do. As many applications could benefit from forecasting an entity's behavior for decision making, assessment, and training, it has gained increasing attention in recent years. Its applications range from homeland and national security to government policy evaluation and market analysis to pandemic and disaster response planning, to name a few. Research on group behavior prediction centers on building predictive models for socio-cultural-political modeling, which includes three key issues: data collection, model construction, and forecasting using the model [1].

Group behavior prediction provides an ideal testbed for practicing and evaluating plan inference approaches. There are huge amounts of group data available online. Recent progress has made it possible to automatically extract plan knowledge (i.e. actions, their preconditions, and effects) from online raw textual data and construct group plans by means of a planning algorithm, albeit in the restrictive security informatics domain [2]. Compared with machine learning-based methods, plan-based inference provides additional advantages for representing, analyzing, and explaining behavior prediction results. Plans are more expressive in representing behavioral patterns, group strategies, and alternative courses of actions, which provide important information for behavioral analysis. With the structural plan representation, plan-based inference will not only come up with prediction results, it can also inform the goals and intentions behind the predicted behavior. Thus, in contrast to data-driven approaches, plan inference can provide richer prediction results and improve interpretability.

In this chapter, we present a probabilistic plan-based approach to forecasting group behavior. Plan representations are typically used by many intelligent systems, especially agent-based systems. Plans provide a concise description of the causal relationship between goals, events, and states. They also provide a clear structure for exploring alternative courses of actions, and interactions between future activities. Such representations have several key advantages: recognizing the relevance of events to agents' goals and plans—

Advances in Intelligence and Security Informatics. DOI: 10.1016/B978-0-12-397200-2.00004-X key for intention recognition; assessing agents' freedom and choice in acting—key for the assessment of power and control; and detecting how an agent's plan facilitates or prevents the plan execution of other agents—key for the detection of intervention. In addition to proposing this computational approach, we conduct an experiment to evaluate the proposed approach in group behavior prediction.

4.1 Review of Plan-Based Inference

Plan inference (or plan recognition) is the process of inferring the plans and goals of the observed agent based on a sequence of observations, usually with the help of a set of predefined recipes (called a plan library), which comprises the knowledge and action steps that can be performed by the agent. Most previous research on plan recognition has focused on the influence of the observed actions on the recognition task. World states, especially the observed agent's preferences over states, are often ignored in the recognition process. In our proposed approach, we view plan recognition as inferring the decision-making strategy of the observed agent (or a group of agents) and explicitly take the observed agent's preferences into consideration.

Previous research identifies the properties of intentions in practical reasoning, and states intentions as elements in agents' stable partial plans of action structuring present and future conduct. Plans thus provide context in inferring intentions, pertaining to the goals of and reasons for an agent's behavior. This justifies plan inference as a means to recognize the intentions and goals of an agent. Utility and rationality issues are also explored in AI and agent research, as a means for specifying, designing, and controlling rational behavior, as well as a descriptive means for understanding behavior. In our approach, we use utilities to represent the presumed preferences of the observed agents. State preferences are used in recognizing the intentions of agents and for disambiguation.

Meanwhile, in AI literature, there is a wealth of computational work on plan/intention recognition. Due to limitations of space, here we only list the most relevant work. Charniak and Goldman [3] proposed the first probabilistic model to deal with the uncertainty inherent in plan inference. Their model is based on Bayesian reasoning. Huber et al. [4] use PRS as a general specification language, and construct the dynamic mapping from PRS to belief networks for plan recognition. Because of the similarity between plan recognition and natural language parsing, Pynadath and Wellman [5] proposed a probabilistic reasoning method based on probabilistic state-dependent grammars (PSDGs). Bui et al. [6] proposed an online probabilistic policy recognition method based on the abstract hidden Markov model (AHMM). More recently, Avrahami-Zilberbrand and Kaminka [7] presented a hybrid approach that combines a symbolic plan recognizer with a decision-theoretic inference mechanism to capture the observer's own biases and preferences. Geib and Goldman [8] presented a probabilistic plan recognition algorithm based on a plan execution model.

Though the approaches differ, most plan recognition models infer a hypothesized plan from observation of actions. World states and, in particular, the observed agents' preferences over outcomes are often ignored in the recognition process. On the other hand, in many real-world applications, the utility of different outcomes is clearly known. Therefore, when a planning agent makes decisions and acts in a real-world situation, it needs to explicitly take this information into account and balance between different possible outcomes. In this chapter, we propose a decision-theoretic approach to plan recognition and explicitly model the observed entity's state preferences in the recognition process. We also conduct an experimental study to validate our approach based on group plans and preferences in real-world scenarios. Based on the realistic group data, we construct a plan library and compare the predictions of our approach with those of an alternative approach with respect to human predictions.

4.2 Probabilistic Plan Representation

Plan representations are used by many intelligent systems. In a traditional plan representation, an action A has preconditions and effects. The action *precondition* is the state that must be made true before action execution. The action *effect* (including *conditional effect*) is the state achieved after action execution. *Antecedents* and *consequences* of conditional effects are also world states. If the antecedents of a conditional effect hold before action execution, its consequences will likely hold after action execution. Actions can be either *primitive* (i.e. directly executable by agents) or *abstract*.

In a probabilistic plan representation, the likelihood of states is represented by probability values. To represent the success and failure of action execution, we use *execution probability* $P_{execution}$ to represent the likelihood of successful action execution given that action preconditions are true. An action effect can be nondeterministic and/or conditional nondeterministic. We use *effect probability* P_{effect} to represent the likelihood of the occurrence of an action effect given that the corresponding action is successfully executed, and *conditional probability* $P_{conditional}$ to represent the likelihood of the occurrence of its consequence given that a conditional effect and its antecedents are true. The desirability of action effects (i.e. their positive/negative significance to an agent) is represented by *utility* values. *Outcomes* are those action effects with nonzero utility values.

A plan P is a partially ordered action sequence to achieve certain intended goal(s). In a probabilistic plan representation, we use expected utility (EU) to represent the overall benefit or disadvantage of a plan.

4.3 Probabilistic Reasoning Approach

Our approach is based on the fundamental *MEU* ('maximum expected utility') principle underlying decision theory, which assumes that a rational agent will adopt a plan maximizing the expected utility. The computation of expected plan utility captures two important factors: one is the desirability of plan outcomes, the other is the likelihood of outcome occurrence, represented as outcome probability. The calculation of outcome probability considers three sources of uncertainty: uncertainty in action preconditions (i.e. state probabilities), uncertainty in action execution (i.e. execution probabilities), and nondeterministic and/or conditional action effects (i.e. effect probabilities). Before presenting our computational approach, we first introduce the notation we adopt.

4.3.1 Notation

Let E be the evidence. Let A, e, c, o, and P be an action, an action effect, a consequence of a conditional effect, an outcome, and a plan respectively. The following notation is adopted in our approach:

- precondition(A): precondition set of action A.
- effect(A): effect set of action A.
- conditional effect(A): conditional effect set of action A.
- antecedent(e): antecedent set of conditional effect e.
- consequence(e): consequence set of conditional effect e.
- $P_{effect}(e|A)$: probability of the occurrence of its effect e given action A is successfully executed.
- $P_{conditional}(c|\text{antecedent}(e), e)$: probability of the occurrence of its consequence c given conditional effect e and its antecedents are true.
- $P_{execution}(A|\text{precondition}(A))$: probability of successful execution of action A given its preconditions are true.
- $P_{action}(o|E)$: probability of action outcome o given evidence E.
- $P_{plan}(o|E)$: probability of plan outcome o given evidence E.
- utility(o): utility value of outcome o (ranging between -100 and $+100$ in the model).
- $EU(A|E)$: expected utility of action A given evidence E.
- $EU(P|E)$: expected utility of plan P given evidence E.

4.3.2 Computation

The computation of expected plan utility is similar to that in decision-theoretic planning, using the utilities of outcomes and the probabilities with which different outcomes occur. In our approach, however, we use the observed evidence to incrementally update state probabilities and the probabilities of action execution, and compute an exact utility value rather than a range of utility values as in decision-theoretic planning. The computation process is realized through recursively using plan knowledge represented in plans.

Probability of States

Let E be the evidence. If state x is observed, the probability of x given E is 1.0. Observations of actions change the probabilities of states. If action A is observed to be executed, the probability of each precondition of A should be 1.0, and the probability of each effect of A is its execution probability multiplied by the effect probability. If A has conditional effects, the probability of a consequence of a conditional effect of A is the product of its execution probability, conditional probability, and the probabilities of each antecedent of the conditional effect:

- If $x \in$ precondition(A), $P(x|E) = 1.0$.
- If $x \in$ effect(A), $P(x|E) = P_{\text{execution}}(A|\text{precondition}(A)) \times P_{\text{effect}}(x|A)$.
- If $x \in$ consequence(e) $\wedge$ $e \in$ conditional effect(A),

$$P(x \mid E) = P_{\text{execution}}(A \mid \text{precondition}(A)) \times P_{\text{conditional}}(x \mid \text{antecedent}(e), e) \times \prod_{e' \in \text{antecedent}(e)} P(e' \mid E)$$

If an action A is observed to be executed, given this evidence, the probability of executing A is 1.0. In this case, the computation above can be simplified:

- If $x \in$ precondition(A), $P(x|E) = 1.0$.
- If $x \in$ effect(A), $P(x|E) = P_{\text{effect}}(x|A)$.
- If $x \in$ consequence(e) $\wedge$ $e \in$ conditional effect(A),

$$P(x \mid E) = P_{\text{conditional}}(x \mid \text{antecedent}(e), e) \times \prod_{e' \in \text{antecedent}(e)} P(e' \mid E)$$

Otherwise, the probability of x given E is equal to the prior probability of x.

Probability of Action Execution

If an action A is observed to be executed, the probability of successful execution of A given E is 1.0, i.e. $P(A|E) = 1.0$. If A is observed to be executed, $P(A|E)$ equals its execution probability. Otherwise, the probability of successful execution of A given E is computed by multiplying the execution probability of A and the probabilities of each action precondition:

$$P(A \mid E) = P_{\text{execution}}(A \mid \text{precondition}(A)) \times \prod_{e \in \text{precondition}(A)} P(e \mid E)$$

So the changes in state probabilities affect the probability calculation of action preconditions, and the probabilities of action execution are changed accordingly.

Outcome Probability and Expected Utility of Actions

Changes of probability in action execution affect the calculation of outcome probabilities and expected utilities of actions. Let O_A be the outcome set of action A, and outcome $o_i \in O_A$. The probability of o_i given E is computed by multiplying the probability of executing A and the effect probability of o_i:

$$P_{action}(o_i \mid E) = P(A \mid E) \times P_{effect}(o_i \mid A)$$

If o_i is the consequence of conditional effect e of A, the formula above should also include the probabilities of each antecedent of the conditional effect:

$$P_{action}(o_i \mid E) = P(A \mid E) \times P_{conditional}(o_i \mid antecedent(e), e) \times \prod_{e' \in antecedent(e)} P(e' \mid E)$$

The expected utility of A given E is computed using the utilities of each action's outcome in A and the probabilities with which each outcome occurs:

$$EU(A \mid E) = \sum_{o_i \in O_A} (P_{action}(o_i \mid E) \times \text{utility}(o_i))$$

Outcome Probability of Plans and Expected Plan Utility

Similarly, changes in the probability of action execution affect the calculation of outcome probabilities and expected plan utilities. Let O_P be the outcome set of plan P, and outcome $o_j \in O_P$. Let $\{A_1, \ldots, A_k\}$ be the partially ordered action set in P leading to o_j, where o_j is an action effect of A_k. The probability of o_j given E is computed by multiplying the probabilities of executing each action leading to o_j and the effect probability of o_j (note that $P(A_i|E)$ is computed according to the partial order of A_i in P):

$$P_{plan}(o_j \mid E) = \left(\prod_{i=1,\ldots,k} P(A_i \mid E) \right) \times P_{effect}(o_j \mid A_k)$$

If o_j is the consequence of conditional effect e of A_k, the formula above should also include the probabilities of each antecedent of the conditional effect:

$$P_{plan}(o_j \mid E) = \left(\prod_{i=1,\ldots,k} P(A_i \mid E) \right) \times P_{conditional}(o_j \mid antecedent(e), e) \times \left(\prod_{e' \in antecedent(e)} P(e' \mid E) \right)$$

The expected utility of P given E is computed using the utilities of each plan outcome in P and the probabilities with which each outcome occurs:

$$EU(P \mid E) = \sum_{o_j \in O_P} (P_{plan}(o_j \mid E) \times \text{utility}(o_j))$$

The intention recognition algorithm works on a possible plan set that is a subset of the plan library. Each plan in the possible plan set includes some or all of the observed actions/ states. Observations are processed one by one, and the probabilities of corresponding actions and states are updated accordingly. Finally, the algorithm calculates the expected utilities of each possible plan; the one with the highest expected utility is inferred as the current hypothesized plan.

4.4　Case Study in Security Informatics

We evaluate the effectiveness of our approach using a three-step procedure. First, as intention recognition relies on a plan library indicating plan knowledge and recipes of plans, we construct a domain plan library using online group data. Second, based on the randomly generated evidence set and the plan library, human raters help build the test set by providing predictions associated with each line of evidence. Finally, to validate our approach, we compare our model predictions with the prediction results using Bayesian reasoning against human predictions.

4.4.1　Construction of Plan Library

We conducted our experiment in the security informatics domain and chose Al-Qaeda as a representative radical group for our study. Group plans can be written manually by domain experts. However, due to the workload of hand-made plans, inconsistency between different experts, and the complexity of group behavior, this method is impractical and error-prone in practice. As a huge volume of reports about this group and its historical events is available online, we employed computational methods to automatically generate group attack plans from relevant open-source textual data [2].

The textual data we used were news items about Al-Qaeda reported from 2000 to 2009 in *Times Online* and *USA Today*, with a total of 10,419 Web pages. Group actions were acquired by extracting verb–object pairs in each sentence where the subject was the name of the group. We designed a number of linguistic patterns and used syntax parsing to extract knowledge of action preconditions and action effects for the automatic construction of domain theory [2]. The extracted group actions were then refined by unifying syntactic forms, combining semantically similar pairs and eliminating static verbs and low-frequency ones, all referencing the WordNet. The refinements of action preconditions and effects were performed similarly. We collected a total of 503 group actions, 110 action preconditions, and 60 action effects with quality [2].

One major difficulty of domain knowledge extraction is that some commonsense knowledge is seldom mentioned explicitly in online news. For example, the action '*get visa*' has the effect '*have visa*', but this piece of knowledge is difficult to obtain online. We compensated for missing preconditions and effects associated with group actions by adding commonsense knowledge of the verbs in the action description. With the complete domain theory, we then employed a planning algorithm to automatically generate the attack plans of the group [2].

Among the official investigation reports, 13 real attacks perpetrated by Al-Qaeda have relatively complete descriptions. Based on our automatically generated plans, an intelligence analyst helped choose 13 plans that matched the reported real attacks. These plans formed the plan library for our experimental study. Another consideration is that, although using large numbers of plans is computationally feasible with our approach, we would prefer a relatively small and realistic plan library so that it is tractable by human raters in the experiment.

Figure 4.1 shows a group attack plan in the plan library. The corresponding action knowledge, action execution probabilities, effect probabilities, and utilities are also given in the figure. Outcome utilities of the plans are the normalized values calculated based on the GTD (Global Terrorism Database) data of the reported real or estimated damage (cases of successes or failed attempts) of the actual attacks by this group in history (the assumption here is that causing loss or damage is desirable to this group). The average length of the plans in the plan library is 9.8 (including start and end nodes).

4.4.2 The Test Set

We randomly generated a set of evidence using a combination of actions and initial world states in the plan library. We arranged these actions and states into eight classes, in which similar action/state pairs or mutually exclusive actions were grouped together. For example, {Action: *raise fund*; State: *have money*} and {Action: *buy vehicle*; State: *have vehicle*} are similar action/state pairs, belonging to Class 1 and Class 4 respectively; {Action: *buy bomb*; Action: *build bomb*} and {Action: *take plane*; Action: *take train*} are mutually exclusive action sets, belonging to Class 5 and Class 7 respectively. Class 8 contains the last actions in each plan, such as *plane attack, plane bombing, (railroad) train bombing, car bombing, suicide bombing, bomb attack*, and *shoot attack*. In order to obtain a meaningful evidence set, the random generation process only selected actions and/or states from different classes.

However, we did not use Class 8 in generating the evidence set. These last actions in plans are much more indicative (which directly associate with the goals of each plan). Therefore, our approach usually yielded good results when these actions were given. We also deleted the generated evidence with conflict actions or action/state pairs, e.g. *train hijack* and *take (railroad) train*. We collected a total of 95 lines of evidence. Each line contained either two

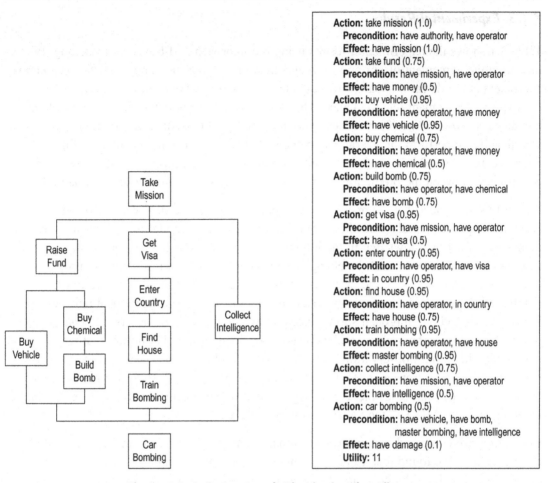

Figure 4.1 A Group Attack Plan in the Plan Library.

observations (constituting 49% of the evidence set) or three observations (constituting 51% of the evidence set).

Four human raters experienced in security informatics participated in the experiment. According to the plan library we constructed, each rater examined the evidence set line by line and predicted the most likely plans based on each line of evidence. The test set is composed of each rater's predictions together with the corresponding evidence, with inter-rater agreement (kappa) of 0.764. The previous state probabilities, action execution probabilities, and effect probabilities used by our approach (less than 100 items in total) were assigned by an intelligence analyst. The intelligence analyst also assigned previous and conditional probabilities for Bayesian reasoning. Mapping plans to Bayesian networks is based on the generic method provided in Ref. [4].

4.4.3 Experimental Results

Table 4.1 shows the experimental results using our approach and Bayesian reasoning. We measured the agreement of our approach and each rater using the kappa statistic. The kappa coefficient is the de facto standard to evaluate the agreement between raters, which factors out expected agreement due to chance. The average agreements between our approach and human raters were 0.664 (for two observations) and 0.773 (for three observations), which significantly outperform the average agreements between Bayesian reasoning and the raters. As $0.6 < \kappa < 0.8$ indicates substantial agreement, the empirical results show good consistency between the predictions generated by our approach and those of human raters.

The results also show that, compared to Bayesian reasoning, the performance of our approach improves rapidly with the increase in the amount of evidence. As our approach makes use of action knowledge in the inference process, actions and states are closely associated in computation and the change in one action or state will be quickly propagated to other interrelated actions and/or states. Thus, our approach is more sensitive to the number of observations. For four observations, there is unanimous agreement between human raters, and our approach shows excellent agreement with the raters (i.e. *convergence point*). As our algorithm only considers a subset of plans in the plan library that is consistent with current observations (i.e. each possible plan being considered includes at least one observed action or state), for a fair comparison we apply Bayesian reasoning in the same way (this increases the average agreement of Bayesian reasoning and human raters from an average of 0.25 to the values given in Table 4.1).

In addition, we compared the answers of human raters with the *1-best* and *2-best* results of the algorithms. We found that a high percentage of the *2-best* results generated by our algorithm fell within the raters' answers (see Table 4.2). Bayesian reasoning also improves considerably in the *2-best* case. The average percentages of the total match for our approach were 75% (for *1-best*) and 90.26% (for *2-best*).

Table 4.1 Kappa Agreements between Algorithms and Human Raters

Rater	Plan Inference						Bayesian Reasoning					
	Two Observations			Three Observations			Two Observations			Three Observations		
	P(A)	P(E)	κ	P(A)	P(E)	κ	P(A)	P(E)	κ	P(A)	P(E)	κ
1	0.826	0.108	0.805	0.878	0.106	0.864	0.436	0.078	0.388	0.469	0.106	0.406
2	0.696	0.090	0.666	0.837	0.105	0.818	0.435	0.086	0.382	0.469	0.107	0.405
3	0.609	0.096	0.567	0.776	0.097	0.752	0.435	0.098	0.374	0.490	0.102	0.432
4	0.652	0.088	0.618	0.694	0.101	0.660	0.370	0.089	0.308	0.388	0.092	0.326
Ave.			0.664			0.773			0.363			0.392

Table 4.2 Comparison of Algorithms' *1-Best* and *2-Best* Results

Rater	Plan Inference				Bayesian Reasoning			
	1-Best		2-Best		1-Best		2-Best	
	#Match	*#Error*	*#Match*	*#Error*	*#Match*	*#Error*	*#Match*	*#Error*
1	81	14	91	4	43	52	58	37
2	73	22	85	10	43	52	56	39
3	66	29	84	11	44	51	64	31
4	65	30	83	12	36	59	57	38
Percent	**75%**	**25%**	**90.26%**	9.74%	**43.68%**	56.32%	**61.84%**	38.16%

4.5 Conclusion

Group behavior prediction is an emerging research and application field that has attracted increasing attention in recent years. It provides an ideal testbed for practicing and evaluating plan inference approaches. In this chapter, we present a plan-based approach to group behavior forecasting based on the principle of maximizing expected plan utility. The proposed approach considers both actions and states in the recognition process, and explicitly takes the observed agent's preferences into consideration. Online plan recognition is realized by incrementally using plan knowledge and observations to change state probabilities and the probabilities of action/plan execution and outcome achievement. Based on realistic online group data, we construct a plan library and conduct an experimental study to evaluate the approach in predicting group behavior. We empirically compared the results with an alternative probabilistic reasoning approach. The experimental results demonstrate the effectiveness of our proposed approach, as well as its compatibility with human intuitions in intention/goal recognition.

In addition to group behavior prediction, we believe our approach is applicable to a wide range of fields in intelligence and security informatics, such as modeling organization behavior in artificial society [9] and extending intention recognition of entities to facilitate social computing and social intelligence [10].

References

1. V. Martinez, G.I. Simari, A. Sliva, V.S. Subrahmanian, CONVEX: Similarity-based algorithms for forecasting group behavior, IEEE Intelligent Systems 23 (4; July/August) (2008) 51–57.
2. X. Li, W. Mao, D. Zeng, F.-Y. Wang, Automatic construction of domain theory for attack planning, Proceedings of the 2010 IEEE International Conference on Intelligence and Security Informatics, 2010, pp. 65–70.
3. E. Charniak, R. Goldman, A Bayesian model of plan recognition, Artificial Intelligence 64 (1), (1993) 53–79.

4. M.J. Huber, E.H. Durfee, M.P. Wellman, The automated mapping of plans for plan recognition, Proceedings of the Tenth Annual Conference on Uncertainty in Artificial Intelligence, 1994.
5. D.V. Pynadath, M.P. Wellman, Probabilistic state-dependent grammars for plan recognition, Proceedings of the Sixteenth Conference on Uncertainty in Artificial Intelligence, 2000.
6. H. Bui, S. Venkatesh, G. West, Policy recognition in the abstract hidden Markov model, Journal of Artificial Intelligence Research 17 (2002) 451–499.
7. D. Avrahami-Zilberbrand, G.A. Kaminka, Incorporating observer biases in keyhole plan recognition, Proceedings of the Twenty-First National Conference on Artificial Intelligence, 2007, pp. 944–949.
8. C.W. Geib, R. Goldman, A probabilistic plan recognition algorithm based on plan tree grammars, Artificial Intelligence 173 (2009) 1101–1132.
9. F.-Y. Wang, The emergence of intelligent enterprises: From CPS to CPSS, IEEE Intelligent Systems 25 (4; July/August) (2010) 85–88.
10. F.-Y. Wang, K.M. Carley, D. Zeng, W. Mao, Social computing: from social informatics to social intelligence, IEEE Intelligent Systems 22 (2; March/April) (2007) 79–83.

CHAPTER 5

Forecasting Complex Group Behavior via Multiple Plan Recognition

In recent years, group behavior forecasting has been attracting attention in a number of ISI research and application fields, such as homeland and national security, government policy evaluation, market analysis, pandemics and disaster response planning, to name just a few. Research on group behavior forecasting has mainly focused on building predictive models based on statistical learning methods [1,2]. Although machine learning methods provide a computational means to construct predictive models, they heavily rely on the structured data, which are usually hard to obtain. Furthermore, because of their data-driven features, the explanation of prediction results given by these methods is also rather weak.

To overcome these limitations inherent in machine learning methods, we take a plan-based approach to group behavior forecasting. A plan-based approach typically relies on a predefined plan library, which comprises the action knowledge and action steps that can be performed by the agent. Recent progress has made it possible to automatically extract plan knowledge (i.e. actions, their preconditions, and effects) from online raw textual data and construct a group plan library by means of a planning algorithm [3]. Compared with machine learning-based methods, plan representation is more expressive in representing the goals, intentions, and alternative courses of actions behind the predicted group behavior.

The plan-based inference and prediction process is called plan recognition (PR), a process of inferring the plans and goals of the observed agent based on a sequence of observations, usually with the help of a plan library. For the reasons mentioned above, plan recognition is particularly suitable for forecasting group behavior. However, previous work on plan recognition often makes a simplifying assumption, i.e. an agent only commits to one plan at a time. In real-world situations, where a group often engages in complex behavior and may pursue multiple plans/goals simultaneously, this assumption is often violated. To infer complex group behavior, new methods capable of recognizing multiple plans are in great demand.

Multiple plan recognition (MPR) can handle cases where a group is carrying out either a single plan or multiple plans. Thus, it is generally applicable in forecasting complex group behavior (single PR can actually be viewed as a special case of MPR). Given the benefits provided by MPR, it poses many more challenges from a computational perspective.

Advances in Intelligence and Security Informatics. DOI: 10.1016/B978-0-12-397200-2.00005-1

Compared with single PR, MPR needs to take into account many new factors, for example the interaction among multiple goals/plans, actions shared by different plans, and the manner of carrying out multiple plans. In addition, given the same observations, the hypothesis space of MPR is much larger than that of single PR and this greatly increases the computational complexity.

In this chapter, we propose a multiple plan recognition approach to complex group behavior forecasting. Our MPR approach constructs a hypothesis space and maps the space into a directed weighted graph. We then convert a multiple plan recognition problem into a classical graph theory problem (i.e. a *directed Steiner tree problem*) and incorporate a graph search algorithm to find the best method to predict group behavior. We also conduct a case study to empirically evaluate the effectiveness of our proposed approach against human predictions.

5.1 Multiple Plan Recognition for Behavior Prediction

Within the research field of group behavior forecasting, predictive models were typically constructed using machine learning methods. Recently, Martinez et al. [1] proposed the CONVEX algorithm for model construction, which is more computationally efficient than previous algorithms. CONVEX is essentially a variant of the kNN method. Although machine learning can automate the construction of predictive models, the structured data used by the learning algorithms have to be manually coded by humans. For example, CONVEX is based on the MAROB datasets, a collection of historical behavioral records of ethnopolitical groups as well as environmental factors associated with group behavior. The datasets were handcrafted by domain experts who examined large volumes of trusted news reports [1]. This process is not only painstaking, but prohibitive for massive data collection.

In plan recognition research, early work has focused on recognizing a single plan. To deal with the uncertainty inherent in plan inference, the majority of previous work has adopted probabilistic reasoning approaches. Charniak and Goldman [4] built the first probabilistic plan recognition model based on Bayesian networks. Huber et al. [5] constructed dynamic mapping from a general specification language (PRS) to belief networks. Besides Bayesian networks, other typical approaches have been based on dynamic belief networks [6], an abstract hidden Markov model [7], natural language parsing [8], and a plan execution model [9]. Many of these methods assume full observability (e.g., Refs. [7,8,10,11]).

Recent research has begun to address multiple plan/goal recognition. Geib and Goldman [10] first pointed out the importance of recognizing multiple plans and extended their previous plan execution model [9] to account for multiple plan recognition. However, their work assumed a fully observable environment and the computational complexity of their approach is at least NP-hard. Another recent work, by Hu and Yang [11], proposed a two-level probabilistic framework for multiple goal recognition. Their work also assumes full observability and adopts a nonhierarchical plan representation. However, complex group activities are rarely fully observable in practice. Taking partial observability into consideration gives the plan recognizer the flexibility to account for the uncertainty existing in behavior observation.

To address these challenges and make plan recognition applicable to complex group behavior forecasting, in this chapter we propose a multiple plan recognition method for partially observable environments. Our approach takes both observed actions and states as inputs. We choose hierarchical plan representation [12] as the typical representation, as it is widely adopted by many intelligent systems in real-world applications. Recognizing hierarchical plans can infer intentions and goals of agents at different abstraction levels, but at the cost of increasing the workload of recognition tasks, since goals at different levels all need to be identified. In contrast, goal recognition is a reduced problem of plan recognition. It only identifies top-level goals while plan recognition aims to recognize plans leading to goal attainments. As plans provide more predictive content of complex group behavior, we will focus on developing a multiple plan recognition method for group behavior forecasting.

Related work on multiple plan/goal recognition has assumed a fully observable environment. In a fully observable environment, the complete plan execution trace provides valuable information to a plan recognizer, including the first action of the plan, negative evidence, and adjacent actions. In a partially observable environment, however, this information no longer exists. Therefore, the multiple plan recognition problem in a partially observable environment is much more challenging. In addition to the observations of actions, world states are also part of the observations. To develop a realistic multiple plan recognizer for predicting group behavior, we will tackle these difficulties in our proposed MPR approach.

5.2 The MPR Problem Definition

In a hierarchical plan representation, an action can be either primitive (i.e. directly executable by an agent) or abstract. An action has preconditions and effects. Preconditions and effects are world states. An abstract action can be decomposed in one or multiple ways, and each decomposition corresponds to a particular method of action execution. To represent the likelihood of a particular method of action execution, action decomposition is associated with probability values.

A hierarchical plan library is a set of hierarchical partial plans. Each partial plan is made up of abstract and/or primitive actions. The actions in the partial plan form a tree-like structure, where an abstract action corresponds to an AND node (i.e. only one decomposition method exists) or an OR node (i.e. multiple decomposition methods exist) in the plan structure. At an AND node, each child is decomposed from its parent with decomposition probability 1. At an OR node, each child is a specialization of its parent. The sum of specialization probabilities for each child is 1.

For each observation, we have either a primitive action or a state. Given a plan library, an explanation SE_i for a single observation O_i is an action sequence starting from a top-level goal G_0 to O_i: $SE_i = \{G_0, SG_1, SG_2, ..., SG_m, O_i\}$, where SG_1, SG_2, ..., and SG_m are a set of abstract actions. There can be multiple explanations for a single observation. An explanation E_j for an observation set $O = \{O_1, O_2, ..., O_n\}$ is defined as $E_j = SE_1 \cup SE_2 \cup ... \cup SE_n$, where SE_i is an explanation for a single observation O_i. If $SE_1 = SE_2 = ... = SE_n$, the explanation E_j corresponds to single plan. Otherwise it corresponds to multiple plans.

We define a multiple plan recognition problem as follows. Given a hierarchical plan library PL and an observation set O, the task of multiple plan recognition is to find the most likely explanation (i.e. best explanation) E_{max} from the explanation set E (a set of all the possible explanations) for O:

$$E_{max} = \arg\max_{E_1 \in E} P(E_1 \mid O)$$

Figure 5.1 shows an example of the partial plan library in the security informatics domain. There are five partial plans in the plan library, where *launch_attack* and

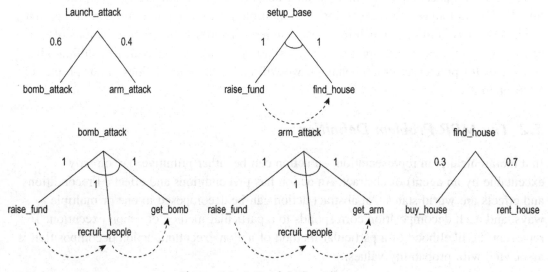

Figure 5.1 A Partial Plan Library.

setup_base are the top-level goals of a group. In the example, *raise_fund*, *recruit_people*, *get_bomb*, *get_arm*, *buy_house*, and *rent_house* are primitive actions, and the other actions are abstract. *Launch_attack* corresponds to an OR node in the plan structure, i.e. the group can either conduct a bomb attack (i.e., *bomb_attack*) with a probability of 0.6 or attack the target by arms (i.e. *arm_attack*) with a probability of 0.4. *Setup_base* corresponds to an AND node, i.e. the group must both raise funds (i.e. *raise_fund*) and then find houses (i.e. *find_house*).

Suppose now we observe two actions that have been executed, *raise_fund* and *buy_house*. We may find several explanations for *raise_fund*: SE_{11} = {*launch_attack, bomb_attack, raise_fund*}, SE_{12} = {*launch_attack, arm_attack, raise_fund*}, and SE_{13} = {*setup_base, raise_fund*}; and an explanation for *buy_house*: SE_2 = {*setup_base, find_house, buy_house*}. Possible explanations for *raise_fund* and *buy_house* include $E_1 = SE_{11} \cup SE_2$, $E_2 = SE_{12} \cup SE_2$, and $E_3 = SE_{13} \cup SE_2$.

5.3 The Proposed MPR Approach

In this section, we propose a probabilistic approach for recognizing multiple plans. As in previous analysis, it is impractical to enumerate all the hypotheses because hypothesis space is often too large. Therefore, we consider using searching techniques to find the best explanation efficiently. Intuitively, if we view the actions of hierarchical plans as vertices and decomposition links as edges, we can convert hierarchical plans into a graph in graph theory. We intend to map multiple plan recognition into a graph theory problem and adopt graph search techniques to find the nearly best explanation.

We first give the definition of the multiple plan recognition problem for group behavior forecasting. Based on this definition, we represent the hypothesis space of input observations as a directed graph (we call it an *explanation graph*). Any possible explanation for the observations is a subgraph of the explanation graph. Next we describe how to compute the probability of an explanation. We then assign weight to the edges of the explanation graph. Finally, we present an algorithm for finding the best explanation from the extended graph and analyze the computational complexity of our approach.

5.3.1 Constructing the Explanation Graph

Here we discuss how to construct an explanation graph for different types of observations (actions or states).

Observed Actions

Given a set of observed actions and a plan library, we first initiate the explanation graph *EG* as an empty graph. We add all the observations to the bottom level of *EG*. We expand the parents of each observation following a breadth-first strategy and add these parents into *EG*. Decomposition/specialization links between actions are treated as directed edges and added to *EG* as well. The direction of the edges denotes decomposition or a specialization relationship. A decomposition/specialization probability is attached to each edge. Duplicate actions and edges are combined during expansion. We continue applying this breadth-first expansion strategy on *EG* until all the actions in *EG* are expanded. We then add a dummy node on the top of the graph and connect the dummy node to all the top-level goals. The edges from dummy node to top-level goals are associated with the previous probabilities of each top-level goal. Now we get an explanation graph that contains all the possible explanations for the given observed actions.

Figure 5.2 shows an explanation graph for the observed actions A_1, A_2, and A_3. It is a directed graph with bold lines denoting an explanation. An explanation corresponds to a connected subgraph in the explanation graph containing dummy node, top-level goals, subgoals, and all the observations. In the explanation, the nodes with input degree 1 correspond to observations.

We notice that an explanation can be either a graph or a tree. (1) If each action in the explanation contributes only to a single goal/subgoal (i.e. each node corresponds to a parent

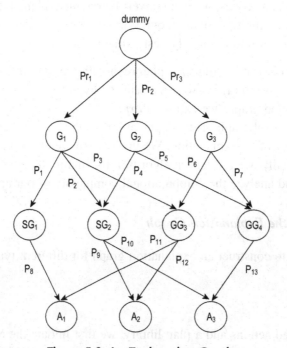

Figure 5.2 An Explanation Graph.

node), the explanation is a tree. (2) If an action in the explanation can contribute to multiple goals/subgoals simultaneously (i.e. a node has multiple parents), the explanation is a graph. An explanation with graph structure implies a specific type of intelligent behavior when an agent executes an action to achieve multiple goals/subgoals simultaneously. As it is relatively less common, we focus on recognizing explanations with a tree structure. For explanations with a graph structure, we will discuss this in Section 5.3.5 and show that this type of explanation can be transferred to a tree structure as well.

Herein we define an explanation for an observation set as a tree in the explanation graph, in which the root is a dummy node and the leaves are all the observations. The tree exactly specifies an explanation for each observation.

Observed States

In partially observable environments, it is a distinct possibility that only the executed action's effects will be observed rather than the action itself. The information about these observed states can help infer unobserved actions and further track an agent's intentions better. We call actions that may cause the observed states candidate actions.

There are three cases of state observation. (1) If the state corresponds to some observed action, we do not treat the state as an observation. (2) If the state corresponds to a new action, we add this action into the observation. (3) If the state corresponds to multiple new actions, we need to infer which action is most likely. Here we use the strategy of maximizing the likelihood parameter estimation. For each candidate action, we add it into the observation set and compute the best explanation. We select the candidate action that maximizes the probability of the best explanation.

To implement the strategy in the explanation graph, we add all the candidate actions (e.g. A_4 and A_5) to the explanation graph and update the explanation graph according to our procedure of constructing the explanation graph. We also add the state (e.g. S_0) to the updated graph and add the directed edges from candidate actions to the state S_0. The probability of the edge from each candidate action to the state is set to 1, which means the execution of candidate action will definitely cause the occurrence of the observed state. We then obtain an extended explanation graph, as shown in Figure 5.3. The candidate action in the best explanation region of the extended graph can be inferred as being the most likely (e.g. in Figure 5.3, A_4 is in the best explanation region and thus is the most likely).

5.3.2 Computing Probability of an Explanation

Given a set of observed actions $O_{1:i} = \{O_1, O_2, \ldots, O_i\}$, the probability of an explanation E_j is computed as

$$P(E_j \mid O_{1:i}) = P(E_j, O_{1:i}) \mid P(O_{1:i}) = P(O_{1:i} \mid E_j)P(E_j) \mid P(O_{1:i})$$

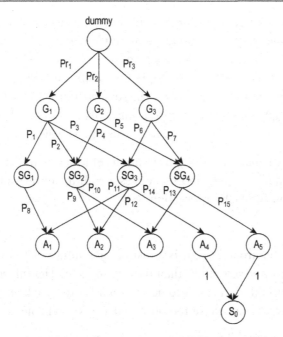

Figure 5.3 The Explanation Graph with Observed State.

As $1/P(O_{1:i})$ is a constant for each explanation, we denote it as K. $P(O_{1:i})|E_j$ is the conditional probability that $O_{1:i}$ occurs given the explanation E_j and is 1 for all the hypotheses. $P(E_j)$ is the prior probability of explanation, i.e. the probability of a tree. For explanation E_j, let $G_{1:m} = \{G_1, ..., G_m\}$ be top-level goals and $SG_{1:n} = \{SG_1, ..., SG_n\}$ be subgoals. We denote the vertex set of the tree E_j as $V =$ dummy $\cup\, G_{1:m} \cup SG_{1:n} \cup O_{1:i}$. Let $E = \{e_1 =$ dummy $\to G_1, ..., e_s = SG_x \to SG_y, ..., e_t = SG_z \to O_i\}$ be the set of edges of E_j, where $1 \le x, y, z \le n$ and $1 \le s \le t$. Here we assume the decomposition of each action is directly influenced by its parent node. The prior probability of the explanation E_j is

$$P(E_j) = P(V, E) = P(O_1, e_t \mid V/O_i, E/e_t) * P(V/O_i, E/e_t)$$

$$= P(e_t) * P(V/O_i, E/e_t) = \cdots = P(\text{dummy}) * \prod_{\text{edge} \in E} P(\text{edge})$$

$P(O_1, e_t|V/Q_i, E/e_t)$ is the conditional probability that the decomposition rule e_t activates and O_i is decomposed given the tree $(V/O_i, E/e_t)$. It equals $P(e_t = SG_z \to O_i)$ according to the assumption. In addition, $P(\text{edge})$ is the probability of the edge in the explanation and $P(\text{dummy})$ is the prior probability that the observed agent is executing plans. It is a constant for each explanation.

5.3.3 Finding the Best Explanation

Now the problem of finding the best explanation can be formulated as

$$
\begin{aligned}
E_{\max} &= \arg\max_{E_j \in E} P(O_{1:i} \mid E_j) P(E_j) \mid P(O_{1:i}) \\
&= \arg\max_{E_j \in E} \prod_{\text{edge}_i \in E_j} P(\text{edge}_i) \\
&= \arg\max_{E_j \in E} \sum_{\text{edge}_i \in E_j} \ln(P(\text{edge}_i)) \\
&= \arg\min_{E_j \in E} \sum_{\text{edge}_i \in E_j} \ln\left(\frac{1}{P(\text{edge}_i)}\right)
\end{aligned}
$$

where $P(\text{edge}_i)$ is the decomposition probability associated with edge_i. We denote $\ln(1/P(\text{edge}_i))$ as the weight of edge_i. As $0 < P(\text{edge}_i) < 1$, we get $\ln(1/P(\text{edge}_i)) > 0$. For explanation graph EG, we attach the weight $\ln(1/P(e))$ to each edge $e \in EG$ (where $P(e)$ is the probability on edge e) and then we can convert an explanation graph to a directed weighted graph. Now the problem of finding the most likely explanation is reformulated as finding a minimum weight tree in the explanation graph with dummy node as the root and observations as leaf nodes.

Finding a minimum weight tree in a directed graph is known as the *directed Steiner tree problem* in graph theory [13,14]. It is defined as follows: given a directed graph $G = (V,E)$ with weights w ($w \geq 0$) on the edges, a set of terminals $S \subseteq V$, and a root vertex r, find a minimum weight tree T rooted at r, such that all the vertices in S are included in T. A number of algorithms have been developed to solve this problem. In our work, we employ an approximation algorithm proposed by Charikar et al. [13].

5.3.4 Algorithm and Complexity Analysis

We have designed an algorithm to find the best explanation, given a hierarchical plan library and an observation set. The algorithm starts from the observations (actions/states). It extends the explanation graph from bottom up, until the root node 'dummy' is extended. After the construction of the explanation graph, the algorithm then computes the weight of each edge in the graph and applies the directed Steiner tree algorithm. Finally, the algorithm returns a directed minimum weight tree (i.e. the best explanation) as the output.

Algorithm (Plan Library PL, *Observation Set* O)

1. Initiate $EG = \{\ \}$, the set of expansion nodes $S = \{\ \}$, and a set $S' = \{\ \}$
2. FOR each observation $o \in O$
   2.1. IF o is an action THEN

    2.2. Add o to EG and S

       END-IF

    2.3. IF o is a state THEN

    2.4.   Find candidate action set C

    2.5.   IF $C \cap O \neq C$

    2.6.   Add all the candidate actions to EG and S

    2.7.   Add the edges from candidate actions to EG

         END-IF

      END-IF

    END-FOR

3. DO

    3.1. FOR each action $A \in S$

       3.1.1. Find parent(A) from PL

       3.1.2. Add parent(A) into EG and S'

       3.1.3. Add the edges from parent(A) to A into EG

       END-FOR

    3.2. $S \leftarrow S, S \leftarrow \{\}$

    WHILE $S \neq \Phi$

4. Add a dummy node d to EG

5. FOR each goal $g \in EG$

    5.1. Add the edge from d to g to EG

    END-FOR

6. Compute the weight of each edge in EG and apply directed Steiner tree algorithm to EG

7. Return a minimum weight tree T as the best explanation.

Complexity

The time complexity of our approach consists of two parts, constructing the explanation graph and searching for the best explanation. The complexity of constructing the explanation graph is the same as that of breadth-first search with duplicate checking in the worst case. The complexity of breadth-first search for K observations is $O(K * b^d)$, where b is the branching factor and d is the maximum depth of the plan library. The complexity of duplicate checking is $O(N^2)$, where N is the number of actions in the plan library. Therefore, the complexity of constructing the explanation graph is $O(K * b^d) + O(N^2)$.

Searching for the best explanation corresponds to the directed Steiner tree problem. Although the complexity of the directed Steiner tree problem has proved to be NP-hard [15], a number of approximation algorithms that can find the nearly best explanation have been developed. The complexity of the approximation algorithm we adopt [13] is $O(N^i K^{2i})$, where K is the number of observations and i is an approximation ratio for tuning the precision of the algorithm. Using the approximation algorithm, our approach can find the nearly optimal explanation in polynomial time.

5.3.5 Discussion

In contrast to the related work on multiple plan/goal recognition [10,11], our work is advantageous in several aspects. First, in real-world applications environments are rarely fully observable, and information about world states is often available in such applications. Our approach works well in partially observable environments. It also takes state observation into consideration. These situations were largely ignored in the related works. Second, compared to the MPR approach proposed in Ref. [10], which is at least NP-hard in complexity, our proposed method is more efficient in that it can employ approximation algorithms to search the best explanation in polynomial time. Third, compared to the multiple goal recognition approach proposed in Ref. [11], our approach works on a hierarchical plan representation, which is widely adopted in many real-world applications. In addition to proposing a novel MPR method, we have conducted human experiments to empirically validate our work in the security informatics domain.

As we mentioned before, the explanation of observations could be a graph instead of a tree and thus an action in the explanation may correspond to multiple parents (multiple subgoals/ goals). In this case, if we represent multiple parents of an action as a single node and add it to the explanation graph, we can convert this type of explanation from a graph structure to tree a structure and apply directed Steiner tree algorithms to compute the probability of this explanation as well.

One limitation of our approach is that it does not take into account the information of ordering constraints in the plan library. An ordering constraint is the order in which actions are executed in partial plans (e.g. the dotted line in Figure 5.1). The ordering constraints of any hypothesis should be consistent with the sequence of observations. It has been utilized in previous plan recognition approaches [7–10] for eliminating hypothesis space. How to extend our current approach to address ordering constraints requires further investigation.

5.4 Case Study in Security Informatics

To validate our proposed method, we conduct a case study in the security informatics domain. In contrast to typical experiments in plan recognition, our study focuses on testing the effectiveness of our approach based on real human data. In the experiment, we first construct a domain plan library using online group data. Then, based on the randomly generated observations and the plan library, human raters help build the test set by providing their predictions associated with the observations. Finally, we compare the results of our algorithm with human predictions.

5.4.1 Experimental Design

We chose Al-Qaeda as a representative group for our study. Based on our previous work [3], we automatically extracted group actions and constructed group plans from relevant open-source news (e.g. *Times Online* and *USA Today*). Domain experts then helped to connect the hierarchical partial plans in the plan library. The plan library used for this experiment includes 10 top-level goals and 35 primitive and abstract actions (we allow primitive/abstract actions to appear in multiple plans). Although large numbers of plans are computationally feasible using our approach, we preferred a relatively small and realistic plan library so that it was tractable by human raters in the experiment. Figure 5.4 illustrates an example of the plan structure for a top-level goal in the plan library.

We randomly generated a number of observation sets using the combination of primitive actions in the plan library. We collected 90 lines of observation sets in total, each line corresponding to one observation set. Among these, 30 observation sets contain two observations each, 30 contain three observations each, and 30 contain four observations each. Five human raters who have at least 3 years' experience in the security informatics domain participated in the experiment. Based on the constructed plan library, each rater examined the observation sets one by one and predicted the most likely plans (single plan or multiple plans) based on each observation set. The test set is composed of each rater's predictions together with corresponding observations (with inter-rater agreement of 0.88).

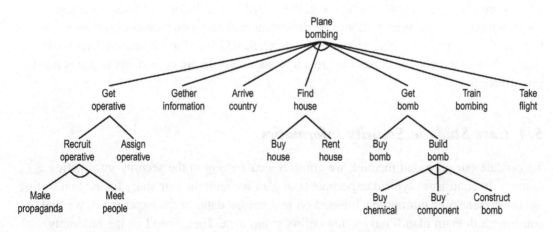

Figure 5.4 Example of the Plan Structure for a Top-Level Goal in the Plan Library.

5.4.2 Results

The computation of the expected plan utility is similar to that in decision-theoretic planning, using the utilities of outcomes and the probabilities with which different outcomes occur. In our approach, however, we use the observed evidence to incrementally update state probabilities and the probabilities of action execution, and compute an exact utility value rather than a range of utility values as in decision-theoretic planning. The computation process is realized through recursively using plan knowledge represented in plans.

Table 5.1 shows the experimental results between our algorithm and each human rater. We measured the average agreement of the results generated by our approach and those of the raters for two, three, and four observations using precision $P(A)$ and kappa statistics κ. The kappa coefficient is the de facto standard to evaluate the agreement between raters, which factors out expected agreement due to chance. The average agreements between our approach and human raters for two observations, three observations, and four observations are all above 0.8. As $0.8 < \kappa < 1$ indicates excellent agreement, the empirical results show good consistency between the predictions generated by our MPR approach and those of human raters.

Table 5.1 Agreements between MPR Algorithm and Human Raters

Rater	Layer	P(A)			κ		
		Two obs.	*Three obs.*	*Four obs.*	*Two obs.*	*Three obs.*	*Four obs.*
#1	0	0.9	0.9	0.883	0.883	0.885	0.870
	1	1.0	0.966	0.922	1.0	0.955	0.903
	2	1.0	0.929	0.893	1.0	0.883	0.848
	Average	**0.967**	**0.932**	**0.899**	**0.961**	**0.908**	**0.874**
#2	0	0.933	0.844	0.842	0.923	0.823	0.824
	1	1.0	0.966	0.902	1.0	0.955	0.879
	2	1.0	0.929	0.862	1.0	0.883	0.808
	Average	**0.978**	**0.913**	**0.869**	**0.974**	**0.887**	**0.837**
#3	0	0.867	0.856	0.758	0.847	0.835	0.736
	1	1.0	0.966	0.922	1.0	0.955	0.902
	2	1.0	0.929	0.897	1.0	0.883	0.853
	Average	**0.956**	**0.917**	**0.859**	**0.949**	**0.891**	**0.830**
#4	0	0.867	0.878	0.892	0.847	0.86	0.88
	1	0.95	0.966	0.922	0.933	0.955	0.903
	2	0.9	0.929	0.893	0.787	0.883	0.848
	Average	**0.906**	**0.924**	**0.902**	**0.856**	**0.899**	**0.877**
#5	0	0.767	0.789	0.867	0.734	0.76	0.852
	1	0.857	0.966	0.904	0.818	0.955	0.882
	2	0.889	0.929	0.862	0.743	0.883	0.808
	Average	**0.838**	**0.895**	**0.878**	**0.765**	**0.866**	**0.847**

Our approach forecasts group behavior in terms of the underlying plans/goals. When a new observation occurs, if it can be explained by a new plan, the new plan together with the original explanations will generate additional explanations and consequently expand the hypothesis space. Therefore, the prediction problem becomes more complex. As shown in Table 5.1, the precision and kappa values tend to decrease with an increase in observations. However, the performance with four observations maintains a considerable level (above 0.8) and thus demonstrates the effectiveness of our approach.

5.5 Conclusion

In this chapter, we present a multiple plan recognition method to forecast the complex behavior of a group. Our MPR method constructs an explanation graph based on the observed actions and/or states, and maps multiple plan recognition into a graph search problem. We have also designed an algorithm to find the best explanation and conducted an experimental study to empirically evaluate our approach against human data. This work demonstrates the effectiveness of our proposed method in forecasting complex group behavior as well as its advantages over related work.

We believe our approach is applicable to a wide range of complex group behavior forecasting applications in the security informatics domain. Our future work will further extend the MPR approach by incorporating the uncertainty in action execution, and handling parameterized plans and online recognition of multiple plans for behavior modeling and prediction.

References

1. V. Martinez, G.I. Simari, A. Sliva, V.S. Subrahmanian, CONVEX: Similarity-based algorithms for forecasting group behavior, IEEE Intelligent Systems 23 (4) (2008) 51–57.
2. S. Khuller, V. Martinez, D. Nau, G. Simari, A. Sliva, V.S. Subrahmanian, Finding most probable worlds of probabilistic logic programs, Proceedings of the First International Conference on Scalable Uncertainty Management, 2007, pp. 45–59.
3. X. Li, W. Mao, D. Zeng, F.-Y. Wang, Automatic construction of domain theory for attack planning, Proceedings of the 2010 IEEE International Conference on Intelligence and Security Informatics, 2010, pp. 65–70.
4. E. Charniak, R. Goldman, A Bayesian model of plan recognition, Artificial Intelligence 64 (1) (1993) 53–79.
5. M.J. Huber, E.H. Durfee, M.P. Wellman, The automated mapping of plans for plan recognition, Proceedings of the Twelfth National Conference on Artificial Intelligence, 1994, pp. 344–351.
6. D.W. Albrecht, I. Zukerman, A.E. Nicholson, Bayesian models for keyhole plan recognition in an adventure game, User Modeling and User-Adapted Interaction 8 (1) (1998) 5–47.
7. H. Bui, S. Venkatesh, G. West, Policy recognition in the abstract hidden Markov model, Journal of Artificial Intelligence Research 17 (1) (2002) 451–499.
8. D.V. Pynadath, M.P. Wellman, Probabilistic state-dependent grammars for plan recognition, Proceedings of the Sixteenth Conference on Uncertainty in Artificial Intelligence, 2000, pp. 507–514.

9. R. Goldman, C. Geib, C. Miller, A new model of plan recognition, Proceedings of the Fifteenth Annual Conference on Uncertainty in Artificial Intelligence, 1999, pp. 245–254.
10. C.W. Geib, R.P. Goldman, A probabilistic plan recognition algorithm based on plan tree grammars, Artificial Intelligence 173 (11) (2009) 1101–1132.
11. D.H. Hu, Q. Yang, CIGAR: Concurrent and interleaving goal and activity recognition, Proceedings of the Twenty-Third National Conference on Artificial Intelligence, 2008, pp. 1363–1368.
12. E. Kutluhan, H. James, S.N. Dana, HTN planning: Complexity and expressivity, Proceedings of the Twelfth National Conference on Artificial Intelligence, 1994, pp. 1123–1128.
13. M. Charikar, C. Chekuri, T. Cheung, Z. Dai, A. Goel, S. Guha, et al., Approximation algorithms for directed Steiner problems, Proceedings of the Ninth Annual ACM-SIAM Symposium on Discrete Algorithms, 1998, pp. 192–200.
14. L. Zosin, S. Khuller, On directed Steiner trees, Proceedings of the Thirteenth Annual ACM-SIAM Symposium on Discrete Algorithms, Society for Industrial and Applied Mathematics, 2002, pp. 59–63.
15. M.R. Garey, D.S. Johnson, Computers and Intractability: A Guide to the Theory of NP-Completeness, W.H. Freeman, 1979.

CHAPTER 6

Social Computing in ISI

A Synthetic View

With the advance of Internet and Web technologies, the increasing accessibility of computer resources and mobile devices, the prevalence of rich media contents, and the ensuing social, economic and cultural changes, computing technology and applications have evolved quickly over the past decade. They now go beyond personal computing, facilitating collaboration and social interactions in general. As such, social computing, a new paradigm of computing and technology development, has become a central theme across a number of information and communication technology (ICT) fields. It has become a hot topic attracting broad interest not only from researchers but also technologists, software and online game vendors, Web entrepreneurs, business strategists, political analysts, and digital government practitioners, to name but a few.

Social computing refers to the computational facilitation of social studies and human social dynamics, as well as the design and use of information and communication technologies that consider the social context. Social computing research methods can help us understand and analyze individual and organizational behavior. In this chapter, we provide a synthetic view of social computing and ISI research, and advocate a viable ISI research framework based on the social computing paradigm. From an information processing perspective, social Web mining techniques can be employed to collect and process data concerning individuals, groups, and their activities. From an analysis and model-building perspective, an emerging social computing research paradigm based on (a) agent-based modeling of artificial societies, (b) computational experiment-enabled scenarios and contingency analysis, and (c) parallel execution-based control and management, can provide a critically needed flexible and comprehensive computational framework for ISI research and development.

6.1 Social Computing

The idea of social computing can be traced back to the 1940s in Vannevar Bush's seminal 1945 *Atlantic Monthly* paper 'As We May Think'. It was not until the 1960s that J.C.R. Licklider headed the Advanced Research Projects Agency (ARPA) and co-wrote *The Computer as a Communication Device* with Robert Taylor. ARPA ultimately led to

Advances in Intelligence and Security Informatics. 10.1016/B978-0-12-397200-2.00006-3

ARPANET, the predecessor of the Internet. Meanwhile, Douglas Englebart's lab at SRI created the first hypermedia online system, NLS (oNLine System). The first collaborative software, EIES (Electronic Information Exchange System), was implemented in the 1970s, and groupware appeared in the 1980s.

Early social software had two distinct foci. One was centered on the technological issues, interfaces, user acceptance, and social effects around group collaboration and online communication. For example, Peter and Trudy Johnson-Lenz defined groupware as 'intentional group processes plus software to support them' [1]. Other definitions of collaborative work and groupware similarly emphasized the group process and supporting software and technologies. The second focus was on the use of computational techniques, principally simulation techniques, to facilitate the study of society and to test out policies before they were employed in real-world organizational or political situations. For example, Richard Cyert and James March utilized simulation to examine how firms behaved [2].

In recent years, the scope of social computing has expanded tremendously, with almost all branches of software research and practice strongly feeling its impact. Table 6.1 lists several recent definitions of social computing and social software. Our definition expands the scope of social computing by including computing technologies that support and help analyze social behavior and help create artificial social agents [3].

6.1.1 Theoretical and Infrastructure Underpinnings

Social computing is a cross-disciplinary research and application field with theoretical underpinnings, including both computational and social sciences (see Figure 6.1). To support social interaction and communication, it relies on: communication; human computer

Table 6.1 Selected Definitions of Social Computing and Social Software

Source	Definition
Communications of the ACM [4]	Describing any type of computing application in which software serves as an intermediary or a focus for a social relation
Wikipedia (http://en.wikipedia.org/wiki)	Referring to the use of social software, a growing trend in ICT usage of tools that support social interaction and communication
Forrester Research [5]	A social structure in which technology gives power to individuals and communities, not institutions
Our definition [6]	Computational facilitation of social studies and human social dynamics, as well as the design and use of ICT technologies that consider social context

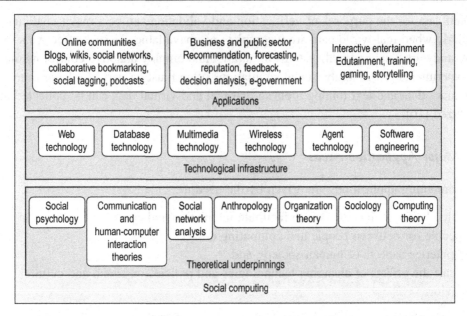

Figure 6.1 Theoretical Underpinnings, Infrastructure, and Applications of Social Computing.

interaction; sociological, psychological, economic, and anthropological theories; and social network analysis [7]. Social informatics studies have revealed that ICT and society influence each other [8]. Thus, social computing has emphasized technology development for society on the one hand and incorporating social theories and practices into ICT development on the other. To facilitate the design of social–technical systems and enhance their performance, social computing must learn from social studies [9] and integrate psychological and organizational theories into computational systems [10]. From an information-processing perspective, the technological infrastructure of social computing encompasses Web, database, multimedia, wireless, agent, and software engineering technologies.

From a methodological viewpoint, incorporating social theories into technology development often poses the additional requirement of constructing artificial societies using agent modeling techniques, according to specific rules and through the interaction of autonomous agents in the environment [11]. Using simulation can be particularly valuable and ethical when examining policies dealing with matters of life and death, such as in bioinformatics, epidemiology [12], and terrorism [13]. In addition, due to the difficulties in testing real systems that are inherently open, dynamic, complex, and unpredictable, computational experiments with artificial systems and simulation techniques are usually needed for evaluating and validating decisions and strategies [14]. Combining real and simulated data for the purposes of verification and validation can be a major challenge, particularly when real-world data are incomplete or unavailable. To seek effective solutions, we can study artificial and real systems in parallel and employ adaptive control methods for these experiments [15]. Finally, social simulations often must be part of a large framework that includes data and text mining tools so that real and virtual data can be collected and co-analyzed [16].

6.1.2 Major Application Areas

Social computing applications are driven by the need to:

- develop better social software to facilitate interaction and communication among groups of people (or between people and computing devices);
- computerize aspects of human society; and
- forecast the effects of changing technologies and policies on social and cultural behavior.

Four main application areas exist. One is the creation of Web services and tools (e.g. blogs, wikis, social networks, RSS, collaborative filtering, and bookmarking) to support effective online communication for social communities. Another application is entertainment software, which focuses on building intelligent entities (programs, agents, or robots) that can interact with human users. Both applications emphasize the technology side and use social theories as guidelines for designing and framing computational systems. The third application area is the business and public sector, which includes various e-businesses, healthcare, economic, political, and digital government systems, as well as artificial engineering systems in domains of significant societal impact. The fourth application area is forecasting, which includes a variety of predictive systems for planning, evaluation, and training in areas ranging from counter-terrorism to market analysis to pandemic and disaster response planning. The last two application areas of social computing are most relevant to security-related applications and play an important role in security informatics.

6.2 A Social Computing-Based ISI Research Framework

Social computing refers to 'the computational facilitation of social studies and human social dynamics as well as the design and use of information and communication technologies that consider social context' [17]. We argue that a social computing-based framework can help us develop a comprehensive ISI research framework and deal with a number of ISI research challenges [18]. In particular, we propose applying an emerging social computing research paradigm called the ACP framework, which is based on (a) agent-based modeling of artificial societies, (b) computational experiment-enabled scenarios and contingency

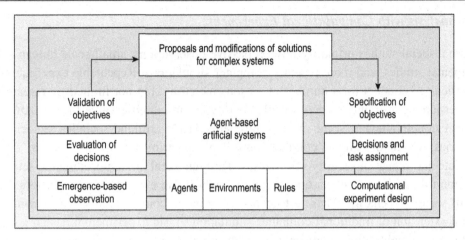

Figure 6.2 A Social Computing-Based ISI Research Framework.

analysis, and (c) parallel execution-based control and management [19]. Figure 6.2 illustrates the ideas of artificial societies, agent-based modeling, and computational experiments that have been applied to the development of artificial social systems in transportation, logistics, and ecosystems [20].

Within the ACP framework, the three major components are inter-related and closely coupled. We first present the research framework for intelligence and security informatics guided by ACP, and then discuss the main research issues involved.

6.2.1 Modeling with Artificial Societies

In the literature, there are no effective formal methods to model complex social–technical systems, especially those heavily involving individual human or group behavior. The ACP framework posits that agent-based artificial societies form the most suitable modeling approach, which can be directly applied to the security informatics domain. An artificial society-based approach has three major components: agents, environments, and rules for interactions. In this modeling approach, how accurately the actual system can be approximated is no longer the only objective of modeling, as is the case in traditional computer simulations. Instead, the artificial society developed is considered as an actual system—an alternative possible realization of the target society. Along this line of thinking, the actual society is also considered as one possible realization. As such, the behaviors of the two societies, the actual and the artificial, are different but fit different evaluation and analysis purposes. Note that approximation with high fidelity is still the desired goal for many applications when it is achievable but can be relaxed otherwise, representing a necessary compromise that recognizes intrinsic limits and constraints of dealing with complex social–technical–behavioral systems in real-world security-related applications.

6.2.2 Analysis with Computational Experiments

Traditional social studies primarily rely on passive observations, small-scale human subject experimental studies, and more recently computer simulations. Repeatable experiments are very difficult to conduct for a number of reasons including, but not limited to, research ethics, resource constraints, uncontrollable conditions, and unobservable factors. Artificial societies can help alleviate some of these problems. Using artificial societies as social laboratories, we can design and conduct controllable experiments that are easy to manipulate and repeat in security informatics. Through agent and environmental setups and interaction rule designs, one can quantitatively analyze and evaluate various factors and 'what-if' security scenarios. These artificial society-based computational experiments are a natural extension of traditional computer simulation. Basic experimental design issues related to model calibration, analysis, and verification need to be addressed. Furthermore, design principles such as replication, randomization, and blocking guide these computational experiments just as they would guide experiments in the physical world.

6.2.3 Control and Management Through Parallel Execution

Parallel execution refers to the fact that long-lived artificial systems can run in parallel and co-evolve with the actual systems they model. This is a generalization of controllers as used in classical automation sciences, which use analytical models to drive targeted physical processes to desired states. This parallel execution idea provides a powerful mechanism for the control and management of complex social systems through co-evolution of actual and artificial systems. The entire system of systems can have three major modes of operation in security informatics. In the 'learning and training' mode, the actual and artificial systems are disconnected. The artificial systems can be used to train personnel. In the 'experimentation and evaluation' mode, connections or synchronizations between the actual and artificial systems take place in discrete time. Computational experiments can be conducted between these synchronizations, evaluating various policies. In the 'control and management' mode, the artificial systems are used as the generalized controllers of the actual systems, with two systems constantly connected. In security informatics applications, those involving control and management of individual/group activities are very important and can benefit directly from the idea of parallel execution.

6.3 Main Issues in the ACP-Based ISI Research Framework

6.3.1 Modeling Cyber-Physical Societies

Behavioral sciences and social psychology provide relevant theory and analysis frameworks to study the influence of social events on psychological and emotional states, at both individual and organization levels. They also help map actions in both cyber and physical spaces to corresponding psychological and emotional states, and security-related events. It is worthwhile investigating the relationships and association between online and offline behavior. Furthermore, social sciences offer root-cause analyses of individual and organizational activities, directly informing the recently emerging adaptive individual/group behavior models using machine learning methods and their predictive applications in security informatics.

As opinion propagation and evolution play a central role in understanding security-related activities and their impact, research that focuses on studying complex structural features of various social networks from the point of view of communication and information sharing is particularly important. Models that take into consideration demographic information and incorporate adaptive spatial–temporal features are urgently called for.

Following the ACP framework, developing an artificial society-based model involves setting up individual/group type, membership distribution, organizational structure, and social networking patterns, based on both the empirical observations through social media, and the social and psychological literature. Furthermore, individual behavior and patterns of interactions with others need to be cataloged and codified. Technically, computational experience of building multi-agent societies, validated at least partially through empirical data, is critical in ISI research. Specific research issues needing further attention are sampled below: How to fuse multi-agent models and complex network models to derive insights that are compatible at both the macro system level and the micro, individual, or small group level. How to develop semantically unified agent models with basic elements to facilitate the engineering aspects of the model building. How to develop a data-driven approach using machine learning and statistical methods to improve individual agent models and interaction protocols in real time with data feeds from security-related social media sources.

6.3.2 Scenario-Based Computational Experiment and Evaluation

A set of computational tools is needed to conduct ISI research to support computational experiments and model evaluation. Automatic or semi-automatic generation of high-fidelity scenarios relevant to the specific events under study is the foundation of computational experiments. Various system-level design choices and parameter configuration and updating schemes need to be evaluated. To conduct computational experiments themselves, an engine to drive discrete-event-based agent simulations, and a platform to plug and play agents for them to interact, are necessary. To support data interpretation and model validation after the experiments, several techniques are relevant: multi-resolution data summarization methods, validated performance evaluation matrices applicable to both empirical data and data generated by computational experiments, and statistical experimental design methods useful for both experimental designs and result evaluation.

Computational experiments and evaluation are particularly useful in two critical areas of ISI research: security-related propagation of ideas, emotions, opinions, and factual information; and evaluation of control and management strategies. The first area has a lot to do with various kinds of 'what-if' network analyses. Based on specific computational experimental scenarios, different types of network topologies can be investigated, given actual hot topic/event evolution and associated information propagation as the backdrop. In addition, simulated hot security-related topics fitting various topic profiles can be experimentally pushed through social networks to observe their emergent behavior. Besides factual information, propagation and dissemination of emotions and opinions can be experimentally explored as well.

In the second area, based on configurable and repeatable computational experiments, one can evaluate and improve security-related control and management policies to either promote or contain certain activities. Note that this experimental method will allow researchers and policymakers to try out various policies that could not be experimented with on a large scale in the real world. Insights learned through these computational experiments are irreplaceable as no other methods (e.g. analytical work or small-scale experiments) can generate these insights with direct applicability to large-scale dynamic social networks.

6.3.3 Interactive Co-Evolution of Artificial and Actual Systems

Research in this area is just emerging and lots of gaps need to be filled. The implications of this notion of parallel execution as part of the ACP framework are far-reaching. To enable computational co-evolution of artificial and actual systems in security informatics, many system-level architectural design tradeoffs need to be considered.

Data representations and communication protocols that support both artificial and actual systems (through sensor networks) need to be formally established. A range of policies governing data synchronization and the interaction mechanism between artificial systems and real systems need to be evaluated. We have discussed three different operational modes of parallel execution in Section 6.2. In the 'control and management' mode, the artificial systems are used as the controller, which can be improved incrementally based on data feeds including feedback to the control policies implemented. In security-related studies, techniques to extract feedback information from social media and associate such feedback signals with policies implemented, and enable semi-automatic adjustments and updates to scenario setups used in computational experiments, are needed. Incremental learning methods useful in a control policy context also need to be developed.

From a policymaking perspective, how to map and associate behavior from artificial systems with real-world observations, how to make sense of emergent behavior observed from co-evolutions of artificial and actual systems, etc., pose additional technical challenges to ISI research.

6.3.4 Social Media Information Processing and Standardization

Social media have become the main platform for radical organizations to conduct business and coordinate their members' activities. Before any in-depth analysis and modeling work can be performed, security-related behavioral data from social media have to be collected and preprocessed. Technically, advances are needed in a number of research areas.

As to information collection, specialized Web-mining techniques that can efficiently extract security-related information from massive and dynamically changing social media, which includes both static Web pages and deep Web components, need to be developed. In terms of information extraction, several techniques are particularly relevant to ISI research: semantic disambiguation of social media contents; user identification and matching across different sites and platforms; entity extraction focusing on security-related events to cover items such as 'who said/discussed what through which channel to whom with what effect'; discovery of relations between users; affect computing techniques that can identify emotional polarity and intensity. After the information extraction step, multiple social networks can be automatically constructed to cover member relationships, event information/emotion propagation, and event correlation. From such networks, identification of opinion leaders, group leaders, and subclusters of users, among others, can be performed. Furthermore, various kinds of system dynamics (e.g. event evolution, membership changes, impact of certain events on group dynamics) can be empirically studied.

As part of the general computational and infrastructural support, we also see a great need for developing ISI application-specific ontologies to facilitate entity extraction and affect computing.

6.3.5 ISI Research Platform

From a research standpoint, it will be vital for the ISI research community to have access to an open research platform to minimize duplicated efforts and to significantly speed up development and evaluation of new research ideas and tools.

We discuss below what we believe are the necessary components of this research platform. Due to the need to process massive security-related social media data and to conduct large-scale multi-agent simulations, this platform needs to be based on a scalable infrastructure in terms of storage and computing power such as the cloud computing infrastructure. It needs to support real-time collection of massive online data and related Web-mining computing needs, and make available Web computing libraries to implement and enable reconfiguration of these Web data collection and mining functions.

For modeling and experimental needs, this platform needs to provide an open and extensible programming environment for modeling and experiments on individual/group behaviors. This environment includes the programming interfaces and bindings to popular programming languages. The Foundation for Intelligent Physical Agents (FIPA) standards and Web services protocols can be effectively utilized to develop agents, and to design agent interaction protocols and interoperable messaging standards. Specialized declarative languages are needed to define security scenarios and computational experimental setups. Furthermore, specialized languages need to be developed to precisely define the interfacing between the artificial and actual systems and enable parallel execution in security informatics applications.

This ISI research platform should also make available a set of tools to facilitate dynamic data visualization, and interactive environmental and experimental designs. In particular, dynamic social network visualization may have multiple applications on this platform.

6.4 Summary

Social computing represents a new computing paradigm and an interdisciplinary research and application field. Undoubtedly, it will strongly influence system and software developments in the years to come. We expect the scope of social computing will continue to expand and its applications in intelligence and security informatics will multiply. From both theoretical and technological perspectives, social computing in ISI research and applications will move beyond social information processing toward emphasizing social intelligence. The move from social informatics to social intelligence can be achieved by modeling and analyzing social behavior, by capturing human social dynamics, and by creating artificial social agents and generating and managing actionable social knowledge in security informatics.

References

1. P. Johnson-Lenz, T. Johnson-Lenz, Consider the groupware: Design and group process impacts on communication in the electronic medium, in: S. Hiltz, E. Kerr (Eds.), Studies of Computer-Mediated Communications Systems: A Synthesis of the Findings, Research Report 16, Computerized Conferencing and Communications Center, New Jersey Institute of Technology, 1981.
2. R. Cyert, J.G. March, A Behavioral Theory of the Firm, Prentice-Hall, 1963.
3. K.M. Carley, A. Newell, The nature of the social agent, Journal of Mathematical Sociology 19 (4) (1994) 221–262.
4. D. Schuler, Social computing, Communications of the ACM 37 (1) (1994) 28–29.

5. C. Charron, J. Favier, C. Li, Social computing: How networks erode institutional power, and what to do about it, Forrester Customer Report, 2006.
6. F.-Y. Wang, Social computing: A digital and dynamical integration of science, technology and human and social studies, China Basic Science 7 (5) (2005) 5–12.
7. S. Wasserman, K. Faust, Social Network Analysis: Methods and Applications, Cambridge University Press, 1994.
8. R. Kling, What is social informatics and why does it matter? D-Lib Magazine 5 (1; January) (1999).
9. M. Douglas, How Institutions Think, Syracuse University Press, 1986.
10. M. Prietula, K.M. Carley, L. Gasser (Eds.), Simulating Organizations: Computational Models of Institutions and Groups, MIT Press, 1998.
11. F.-Y. Wang, J.S. Lansing, From artificial life to artificial societies: New methods for studies of complex social systems, Complex Systems and Complexity Science 1 (1) (2004) 33–41.
12. K.M. Carley, D.B. Fridsma, E. Casman, A. Yahja, N. Altman, L. Chen, et al., BioWar: Scalable agent-based model of bioattacks, IEEE Transactions on Systems, Man, and Cybernetics 36 (2) (2006) 252–265.
13. K.M. Carley, A dynamic network approach to the assessment of terrorist groups and the impact of alternative courses of action, Proceedings of the Visualizing Network Information Meeting, 2006.
14. F.-Y. Wang, Computational experiments for behavior analysis and decision evaluation of complex systems, Journal of System Simulation 16 (5) (2004) 893–897.
15. F.-Y. Wang, Parallel execution methods for management and control of complex systems, Control and Decision 19 (5) (2004) 485–489.
16. K.M. Carley, J. Diesner, J. Reminga, M. Tsvetovat, Toward an interoperable dynamic network analysis toolkit, Decision Support Systems 43 (4) (2007) 1324–1347.
17. F.-Y. Wang, K.M. Carley, D. Zeng, W. Mao, Social computing: From social informatics to social intelligence, IEEE Intelligent Systems 22 (2) (2007) 79–83.
18. F.-Y. Wang, A computational framework for decision analysis and support in ISI: Artificial societies, computational experiments, and parallel systems, Proceedings of the 2006 International Workshop on Intelligence and Security Informatics, 2006, pp. 183–184.
19. F.-Y. Wang, Toward a paradigm shift in social computing: The ACP approach, IEEE Intelligent Systems 22 (5) (2007) 65–67.
20. F.-Y. Wang, S. Tang, Artificial societies for integrated and sustainable development of metropolitan systems, IEEE Intelligent Systems 19 (4) (2004) 82–87.

CHAPTER 7

Cyber-Enabled Social Movement Organizations

The rapid development and wide adoption of Internet and mobile technologies have drastically changed the way people, communities, and organizations communicate and interact. Online media, in particular social media, such as news websites and reader feedback, online newsgroups and forums, chat rooms, blogs, and various social networking sites have been ubiquitously integrated into everyday life. With active participation in the huge Internet and mobile platform (e.g. mobile phones) user base, the increasing socialization of technology use, and intensified continuous interaction between online and offline activities, cyberspace is becoming as real as the physical world, bringing about a fundamental impact on almost all walks of life.

This chapter focuses on an emerging class of organizational actors of major influence in this global cyber-physical space, cyber-enabled social movement organizations. Social movement organizations (SMOs) existed long before the Internet era in an offline context and are traditionally defined as formally organized components of social movements. SMOs play a coordinating role in social movements but do not necessarily employ or direct most of their participants. A cyber-enabled SMO (CeSMO) refers to a group of people, facilitated by the Internet or mobile technologies, that assembles to lead, participate in, discuss, and implement social behavior centered on a defined topic/interest or concerning a particular event. As a special and important type of online community, CeSMOs have recently played a critical role in organizing various movements to promote social change. They are characterized by their volunteerism and activism roots, perplexing self-organizing patterns, sudden and unpredictable onsets, rapid dynamic growth, complex interactivity between online and offline activities, and the potential reach to a large number of users. CeSMOs can cause great social and economic consequences via rapid spread and large-scale diffusion in cyberspace.

CeSMOs and their activities present unprecedented opportunities for researchers in social and computational sciences. CeSMO-related Web contents represent a tremendously rich and useful data source, opening new venues to conduct high-impact data-driven research and empirically test various theories and hypotheses, which has been impossible or prohibitively expensive in the past. In addition, CeSMOs' online and offline behavior is

Advances in Intelligence and Security Informatics. DOI: 10.1016/B978-0-12-397200-2.00007-5

emerging as a subject of scientific investigation and a motivation for new theoretical and empirical research.

7.1 Studies on Social Movement Organizations: A Review

We first review the existing work on social movement organizations, which is a mixture of social and computational studies. Over the last few years, traditional social movements have been drastically affected by information and communication technologies (ICTs), notably the Web and mobile technologies. We use the term cyber-enabled SMOs (CeSMOs) to refer to SMOs in whose activities Internet or mobile technologies play a critical facilitating and enabling role.

Garrett [1] argued that new ICTs are changing the ways in which activists communicate, collaborate, and demonstrate. Choo and Smith [2] considered how technologies can be used by organized crime groups to infringe legal and regulatory controls. They found that the Internet has been explored as a new platform to commit traditional crimes. In addition, new criminal activities have emerged on, and targeted at, this platform. Van Laer and Van Aelst [3] focused on how the Internet has changed the action repertoire of social movements in two fundamental ways. Saeed et al. [4] studied a recent social movement in Pakistan and found that people have started using the Web as a resource for information dissemination and staging online and offline protests. Nah et al. [5] examined how Internet news usage and online political discussions contribute to political participation in anti-war movements. Their analysis revealed that online political discussions mediate certain news media effects on anti-war political participation.

CeSMOs have been studied in crisis management contexts. Hughes et al. [6] outlined several examples of each type of online social convergence behavior during times of crisis: *help*, *be anxious*, *return*, *support*, *mourn*, *exploit*, and *be curious*. Sutton et al. [7] studied information sharing and dissemination practice by the public during the October 2007 Southern California Wildfires. They argued that these emerging uses of social media are precursors of broader future changes to the institutional and organizational arrangements of disaster response. Vieweg et al. [8] analyzed a selected set of online interactions that occurred in the aftermath of the 2007 shooting rampage at Virginia Tech, which represented a new and highly distributed form of participation by the public. These research findings clearly suggest strong self-organization among CeSMOs that includes the development and evolution of roles and norms that guide behavior including, but not limited to, information sharing.

Sociological methods have been widely adopted in CeSMO research. Hollenbeck and Zinkhan [9] investigated the current anti-brand social movement by examining consumer activist groups on the Internet. They found that the Web community reinforces negativity toward the brand and influences a member's attitude or decision to take action against a corporation. Clark and Themudo [10] employed social movement theory to study the influence of dotcauses on social movement's transnational action, leaderlessness, profusion of concerns, tactical schisms, and digital/language divides. Gerstenfeld et al. [11] conducted content analysis on 157 extremist websites. They found that the majority of sites contained external links to other extremist sites. They also found that roughly half of these sites included multimedia content, and that a half contained racist symbols. These and other findings suggest that the Internet has been leveraged as a powerful tool by extremists to reach an international audience, recruit members, link diverse extremist groups together, and conduct public relations campaigns. Zimbra and Chen [12] performed link analyses to examine the relationship between virtual linkage intensity and real-world physical proximity among the social movement groups identified in the Southern Poverty Law Center Spring 2009 Intelligence Report. Their findings indicate the existence of significant relationships between virtual linkage intensity and physical proximity, regardless of ideological categorizations.

There has been much recent interest in employing computational methods to analyze the characteristics of CeSMOs. Data collection is fundamental to these studies. Zhou et al. [13] proposed automated and semi-automated Web-mining procedures and a systematic methodology for capturing Jihad terrorist website data. Chau and Xu [14] proposed a semi-automated approach that combines blog spidering and social network analysis techniques to analyze a selected set of 28 anti-blacks hate groups on Xanga. Community discovery is a hot topic related to CeSMOs. Flake et al. [15] introduced a maximum flow-based Web crawler that can approximate a community by directing a focused Web crawler along highly relevant link paths. Gibson et al. [16] developed a notion of hyperlinked communities on the Web through the analysis of link topology. As a CeSMO implies a network of Internet users, many researchers have adopted social network analysis techniques to explore CeSMO network structure and characteristics. Adamic and Glance [17] studied the linkage patterns and discussion topics of political bloggers using social network analysis. They discovered that liberals and conservatives established links primarily within their separate communities, with far fewer crosslinks cutting across these communities.

Studies of interactions between online and offline activities of CeSMOs are beginning to emerge. Wang et al. [18] argued that human flesh search (HFS) communities can be viewed as a special type of CeSMO and crowdsourcing. They empirically analyzed the HFS events from 2001 to 2010 and suggested that HFS communities featured a much broader range of online and offline interactions than other online communities.

7.2 A New Research Framework for CeSMOs

From a research perspective, CeSMOs pose tremendous challenges to existing ISI research paradigms. The size and real-time dynamics of CeSMOs are often beyond what existing methods can handle. The rich, complex, and unregulated interactions within CeSMOs members and between CeSMOs and the environment, both online and offline, as a new phenomenon, lead to many unexplored research questions calling for new measures and analysis frameworks. The most fundamental among these challenges is the chaotic nature of CeSMOs. Through the amplifying effect of the Web and the mobile networks, CeSMO activities are inherently unpredictable. Small perturbations can lead to major structural changes through real-time self-organization. Also, with the disappearance of most of the resource constraints associated with the physical world, formation and growth of CeSMOs can be fast and volatile.

In this section, we first summarize and catalog CeSMO research questions. A majority of these questions are from the literature, with the remainder based on our own research in this area. We then propose a viable CeSMO research framework based on the social computing paradigm.

7.2.1 CeSMO Research Questions

Table 7.1 presents CeSMO research questions. The first set of questions focuses on the empirical observations of CeSMO formation and evolution, and behavioral and cultural modeling of CeSMOs' behavior and activities using an array of social domain-driven machine learning methods. The second set of questions is concerned with computationally driven identification and discovery of opinion leaders, active members, hot topics, emerging events and prevailing emotions, etc., from CeSMO-generated content, mostly through social media platforms. The third set of questions aims at studying the evolution of CeSMO online communities and the propagation and diffusion of information, opinions, and emotions. The final set of questions is concerned with studying the correlation between CeSMOs' online and offline behavior, and investigating connections between CeSMOs' emotional states and their online and offline behavior.

7.2.2 A Social Computing-Based CeSMO Research Framework

Figure 7.1 illustrates a CeSMO research framework based on the social computing paradigm [19]. Specifically, our proposed CeSMO research framework is based on the ACP methodology. It consists of five major components. The first social media information processing and standardization component focuses on collecting and processing in real or near real time CeSMO-related social media information, and feeding the information to

Table 7.1 CeSMO Research Questions

Categories	Detailed Research Questions
Formation & evolution of CeSMOs, and behavioral & cultural modeling	Formation of CeSMOs Propagation of opinions and emotions in CeSMOs Adaptive behavioral modeling of CeSMOs Social and cultural determinants of CeSMOs CeSMO behavior prediction based on social and cultural information
Identification and discovery	Identifying active CeSMO members and opinion leaders in the social media Identifying relations among CeSMO members in the social media Discovering hot topics/events
Propagation and diffusion of information, ideas, and emotions	Dynamic temporal and spatial models for information propagation and diffusion in CeSMOs Propagation and diffusion of hot topics/events/opinions/emotions in CeSMOs Dissemination of domain events in multi-level social media networks
Online/offline interactions of CeSMO behavior/activities	Induction of social emergencies by CeSMOs behavior Influence of social media information on CeSMOs' psychological and emotional states Mapping between CeSMOs' emotional states and CeSMOs' behavior Correlation between CeSMOs' online and offline behavior

other components. The next three components—the modeling of CeSMOs based on artificial societies covering activities in both cyber (online) and physical (offline) environments; computational experiment-based study and evaluation of CeSMOs; and the interactions between the artificial and actual systems (including both of their cyber and physical components)—are closely mapped to the three key aspects of the ACP method. The final component is concerned with developing a social computing-driven CeSMO research platform, providing a comprehensive set of computational tools and an infrastructure to facilitate CeSMO studies. In each of these components, they share research issues/tasks similar to those identified in Chapter 6. In the next section, we provide a case study of the social computing-based CeSMO research implementing this framework.

7.3 Case Study: Wenchuan Earthquake

We present a case study to illustrate social computing-based CeSMO research. The case study is concerned with information and emotion diffusion during the Wenchuan

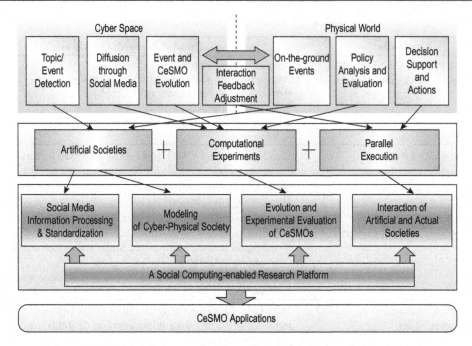

Figure 7.1 A Social Computing-Based CeSMO Research Framework.

earthquake. The Wenchuan earthquake was a huge disaster that happened in 2008. During this disaster, billions of Chinese netizens were concerned about the event and engaged in online dispute and discussion. Rumors and news, different types of emotions, and offline– online interaction were dramatic and the overall picture was mixed, leading to a large-scale complex online phenomenon. We consider it valuable to investigate the dynamics of that cyber system and enhance our understanding of the new phenomenon under the umbrella of social computing. We have constructed reference networks based on online news. Each node in these networks corresponds to a website and each edge indicates that there is at least one Web news citation between the two corresponding nodes.

We collected 78,512 related news items from 1062 websites from 12 May 2008 to 11 June 2008. We are interested in the network structure of news references. Figure 7.2 records the evolution of the news reference network (the total numbers involved are given in Table 7.2). From the figure we can see that the news is distributed rather ordinarily at the beginning. As time goes by, the network displays a core cluster and several small clusters. This may be due to the fact that influential websites have more sources of information and can access the latest news more easily. These sites will quickly become key nodes of the news reference network.

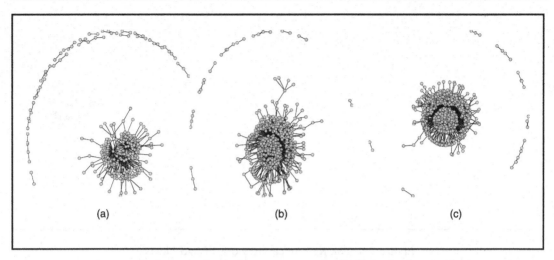

Figure 7.2 The Reference Network of All the News.
(a) Three days after Wenchuan earthquake. (b) One week after Wenchuan earthquake. (c) Two weeks after Wenchuan earthquake.

Table 7.2 The Total Number of News Items

	Period		
	In 3 Days	*In 1 Week*	*In 2 Weeks*
Number of news items	596	10,772	28,287
Number of sites	252	558	839

We adopted clustering algorithms and found six core nodes in this network, i.e. people.com. cn, xinhuanet, tencent, chinanews, sina, and sohu. Among all websites, 66.7% have a reference relationship with these six cores; 64.8% of news is directly related to the cores. This suggests that the news reference network is a core/periphery structure with multiple cores.

Furthermore, we automatically labeled the emotional tendency of news, including sadness, fear, anger, and happiness. Our focus is on discovering the pattern of information propagation, especially emotion propagation. From Figures 7.3–7.6, we can see that fearful news propagates much faster than sad news one week after the Wenchuan earthquake (the totals in each category are recorded in Tables 7.3–7.6). However, reporting of fearful news slows down after two weeks and that of sad news begins to speed up. The main reason may be that a number of rumors dominate in the short term, mostly fearful rumors. However, as more facts are reported, fearful news is less believed by people and sad emotions build up that are diffused over the Internet. We believe that the rate of propagation of news reflects

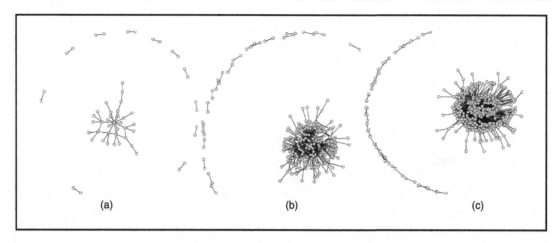

Figure 7.3 The Reference Network of Sad News.
(a) Three days after Wenchuan earthquake. (b) One week after Wenchuan earthquake. (c) Two weeks after Wenchuan earthquake.

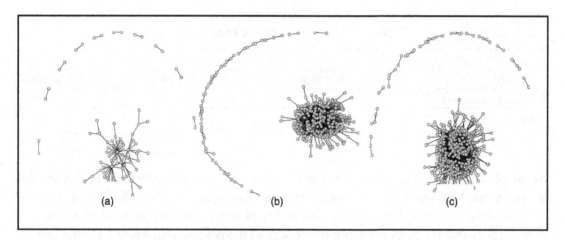

Figure 7.4 The Reference Network of Fearful News.
(a) Three days after Wenchuan earthquake. (b) One week after Wenchuan earthquake. (c) Two weeks after Wenchuan earthquake.

Table 7.3 The Number of Sad News Items

	Period		
	In 3 Days	*In 1 Week*	*In 2 Weeks*
Number of news items	56	753	4396
Number of sites	56	214	524

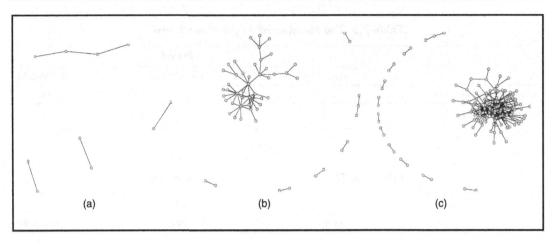

Figure 7.5 The Reference Network of Angry News.
(a) Three days after Wenchuan earthquake. (b) One week after Wenchuan earthquake. (c) Two weeks after Wenchuan earthquake.

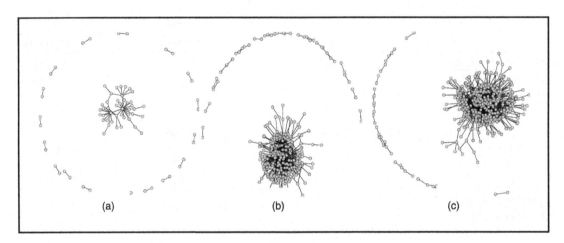

Figure 7.6 The Reference Network of Happy News.
(a) Three days after Wenchuan earthquake. (b) One week after Wenchuan earthquake. (c) Two weeks after Wenchuan earthquake.

Table 7.4 The Number of Fearful News Items

	Period		
	In 3 Days	**In 1 Week**	**In 2 Weeks**
Number of news items	111	1449	3453
Number of sites	79	273	530

Table 7.5 The Number of Angry News Items

	Period		
	In 3 Days	*In 1 Week*	*In 2 Weeks*
Number of news items	6	79	280
Number of sites	10	56	126

Table 7.6 The Number of Happy News Items

	Period		
	In 3 Days	*In 1 Week*	*In 2 Weeks*
Number of news items	78	1672	4393
Number of sites	79	290	483

Table 7.7 The Number of Negative News Items

	Period		
	In 3 Days	*In 1 Week*	*In 2 Weeks*
Number of news items	273	4715	12,225
Number of sites	156	518	646

Table 7.8 The Number of Positive News Items

	Period		
	In 3 Days	*In 1 Week*	*In 2 Weeks*
Number of news items	127	2525	6566
Number of sites	89	334	568

the dynamics of public emotion. If public emotion is consistent with the emotion contained in some news, that news will be shared more and will be propagated more quickly.

We also compute the polarity of news (negative or positive; see Figures 7.7 and 7.8, with the total numbers of each given in Tables 7.7 and 7.8). Not surprisingly, we find negative news propagates much faster than positive news. This confirms our intuition that bad news is propagated more quickly than good news.

We also analyzed the propagation patterns of news on blogs. Figures 7.9 and 7.10 show the propagation patterns of news with different emotions and polarities. We can see that

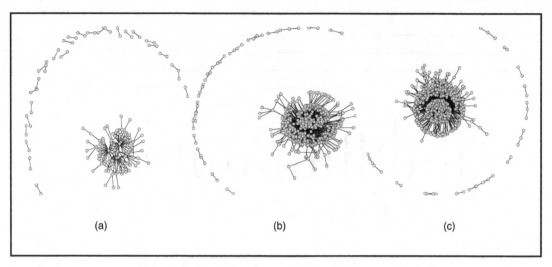

Figure 7.7 The Reference Network of Negative News.
(a) Three days after Wenchuan earthquake. (b) One week after Wenchuan earthquake. (c) Two
weeks after Wenchuan earthquake.

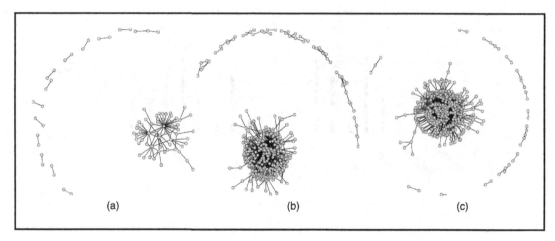

Figure 7.8 The Reference Network of Positive News.
(a) Three days after Wenchuan earthquake. (b) One week after Wenchuan earthquake. (c) Two
weeks after Wenchuan earthquake.

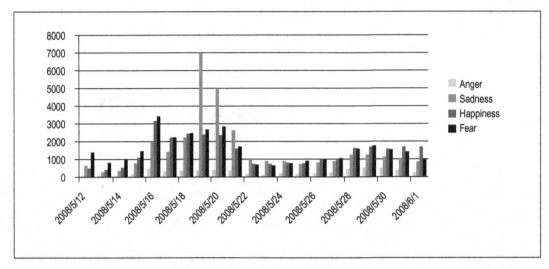

(a) Blogs

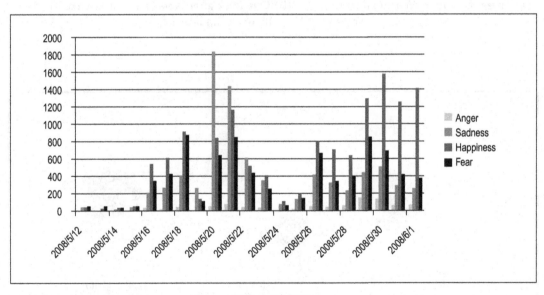

(b) News Web Pages

Figure 7.9 Emotion Evolution on Blogs and News Web Pages.

the trend of emotions on a blog is similar to news Web pages. Fearful emotion was overwhelming several days after the disaster occurred. This implies a pervasive fear of danger among citizens. Sad news was becoming the mainstream as causality data of the earthquake was revealed. Figure 7.10 denotes the evolution of positive and negative news. Negative news dominates positive news at all times on both blogs and news Web

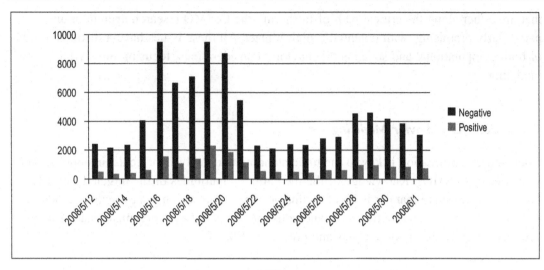

(a) Blogs

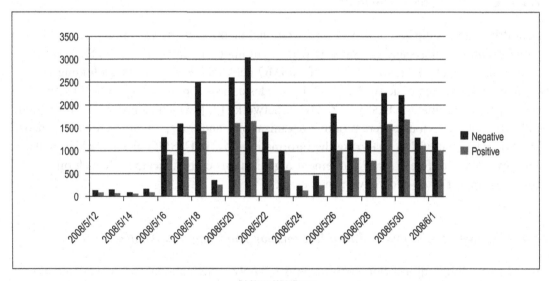

(b) News Web Pages

Figure 7.10 Evolution of Positive and Negative News on Blogs and News Web Pages.

pages. It is more apparent on blogs that negative news is propagated more quickly than positive news.

7.4 Discussions on CeSMO Research Issues

This section presents a number of ongoing CeSMO research issues in the social computing framework. This set of issues was selected to represent the major ongoing research streams that are either along the critical path of furthering the CeSMO research agenda or are particularly promising, with potentially high impact. All these issues have a distinctive mix of both computational and social and behavioral considerations, focusing mostly on modeling.

7.4.1 CeSMO Behavior Modeling

Open-source information is key to the understanding and analysis of goals, intentions, and strategies of CeSMOs. Knowledge of the mentality and attitudes of the targeted CeSMOs in a context-sensitive manner is highly valuable. Specific research issues include modeling CeSMOs' intentions and motivations, and constructing plan-based organizational behavioral patterns for group behavior analysis and prediction [20–22].

7.4.2 CeSMO Network Analysis

Data-driven and computation-driven organizational theories, many of which originated from social network analysis and complex networks, are important in studying the formation, evolution, percolation, and interaction of CeSMO networks. Recent work on finding community structures in networks [23,24] provides a basis for analyzing CeSMO social networks. In addition, the theory of evolving networks [25], which examines the formation and evolution of various different online organizations, and transport and percolation theory [26,27], are also useful in analyzing the information flow in CeSMO groups, which will help reveal how different social movement organizations communicate with each other and how they impact or co-evolve with each other.

7.4.3 CeSMO Social and Cultural Information Modeling and Analysis

Social and cultural information analysis aims to apply computational techniques to build behavioral models of CeSMOs regarding their real-world activities so as to analyze the impact of social and cultural factors on organizational behavior. To this end, core research issues include the automatic construction of group behavior models based on relevant cultural and social factor data items. As input to social and cultural information analysis, the social/cultural attributes cover various social, cultural, religious, economic, or political factors associated with CeSMOs. Machine learning methods such as associative classification and probabilistic inductive logic programming are useful in learning organizational behavior models [28]. Meanwhile, computational problems such as data imbalance and sparsity need to be properly dealt with using various effective data preprocessing mechanisms such as resampling.

7.4.4 CeSMO Behavior Prediction

The behavior of CeSMOs is often greatly influenced by various environmental (e.g. social, political, cultural, financial, and economic) factors. Mining actionable rules for influencing entity behavior has wide applicability in various domains. Actionable behavior rules provide the user with explicit suggestions of actions to influence the behavior of the entity with regard to the user's benefit. This kind of actionable knowledge is important as it can lead to an act, make things happen, and thus one can act upon it [29]. Actionability is also an important aspect that has to be quantified based on the subjectivity of the user to facilitate the decision-making process [29]. Action rule mining can facilitate CeSMO behavior prediction by providing valuable insights for decision makers to estimate the impact of CeSMO behavior and related social movements.

7.5 Conclusion

This chapter focuses on the cyber-enabled SMOs, the study of which poses fundamental methodological challenges to ISI research. We advocate a viable CeSMO research framework based on an emerging social computing research methodology, discuss specific research issues within this methodological framework, and present a research case study to illustrate the application of this framework.

As Internet and mobile technologies continue to penetrate all walks of life globally, CeSMOs, as a research subject, will become increasingly important. From a research perspective, they challenge conventional methods and wisdom, and have become a fruitful base for new ideas and findings, especially from the point of view of cross-disciplinary research. On the application side, CeSMO studies can have major implications in multiple areas relating to intelligence and security informatics, such as politics, national and international security, technology policy, and economics, to name just a few.

References

1. R. Garrett, Protest in an information society: A review of literature on social movements and new ICTs, Information, Communication and Society 9 (2) (2006) 202–224.
2. K. Choo, R. Smith, Criminal exploitation of online systems by organised crime groups, Asian Journal of Criminology 3 (1) (2008) 37–59.
3. J. Van Laer, P. Van Aelst, Cyber-protest and civil society: The Internet and action repertoires in social movements, in: Y. Jewkes, M. Yar (Eds.), Handbook on Internet Crime, Cullompton: Willan, 2009, pp. 230–254.
4. S. Saeed, M. Rohde, V. Wulf, ICTs, an alternative sphere for social movements in Pakistan: A research framework, IADIS International Conference on e-Society, 2008.
5. S. Nah, A.S. Veenstra, D.V. Shah, The Internet and anti-war activism: A case study of information, expression and action, Journal of Computer-Mediated Communication 12 (1) (2006) 230–247.

6. A.L. Hughes, L. Palen, J. Sutton, S.B. Liu, S. Vieweg, "Site-seeing" in disaster: An examination of on-line social convergence, Proceedings of the Fifth International ISCRAM Conference, 2008.

7. J. Sutton, L. Palen, I. Shklovski, Backchannels on the front lines: Emergent uses of social media in the 2007 southern California wildfires, Proceedings of the Fifth International ISCRAM Conference, 2008.

8. S. Vieweg, L. Palen, S. Liu, A. Hughes, J. Sutton, Collective intelligence in disaster: Examination of the phenomenon in the aftermath of the 2007 Virginia Tech shooting, Proceedings of the Fifth International ISCRAM Conference, 2008.

9. C. Hollenbeck, G. Zinkhan, Consumer activism on the Internet: The role of anti-brand communities, Advances in Consumer Research 33 (2006) 479–485.

10. J.D. Clark, N.S. Themudo, Linking the web and the street: Internet-based "dotcauses" and the "anti-globalization" movement, World Development 34 (1) (2006) 50–74.

11. P.B. Gerstenfeld, D.R. Grant, C. Chiang, Hate online: A content analysis of extremist Internet sites, Analyses of Social Issues and Public Policy 3 (1) (2003) 29–44.

12. D. Zimbra, H. Chen, Comparing the virtual linkage intensity and real world proximity of social movements, Proceedings of the 2010 IEEE International Conference on Intelligence and Security Informatics, 2010, pp. 144–146.

13. Y. Zhou, E. Reid, J. Qin, H. Chen, G. Lai, US domestic extremist groups on the Web: Link and content analysis, IEEE Intelligent Systems 20 (5) (2005) 44–51.

14. M. Chau, J. Xu, Mining communities and their relationships in blogs: A study of online hate groups, International Journal of Human–Computer Studies 65 (1) (2007) 57–70.

15. G.W. Flake, S. Lawrence, C.L. Giles, Efficient identification of Web communities, Proceedings of the Sixth ACM SIGKDD International Conference on Knowledge Discovery and Data Mining, 2000, pp. 150–160.

16. D. Gibson, J. Kleinberg, P. Raghavan, Inferring Web communities from link topology, Proceedings of the Ninth ACM Conference on Hypertext and Hypermedia, 1998.

17. L.A. Adamic, N. Glance, The political blogosphere and the 2004 U.S. election: Divided they blog, Proceedings of the Third International Workshop on Link Discovery, 2005.

18. F.-Y. Wang, D. Zeng, J.A. Hendler, Q. Zhang, Z. Feng, Y. Gao, H. Wang, G. Lai, A study of the human flesh search engine: crowd-powered expansion of online knowledge, IEEE Computer 13 (8) (2010) 1–9.

19. F.-Y. Wang, A study of cyber-enabled social movement organizations based on social computing and parallel systems, J. University of Shanghai for Science and Technology 33 (1) (2011) 8–17.

20. K.M. Carley, Computational modeling for reasoning about the social behavior of humans, Computational and Mathematical Organization Theory 15 (1) (2009) 47–59.

21. K.M. Carley, D.B. Fridsma, E. Casman, A. Yahja, N. Altman, L. Chen, B. Kaminsky, D. Nave, BioWar: Scalable agent-based model of bioattacks, IEEE Transactions on Systems, Man, and Cybernetics, Part A: Systems and Humans 36 (2) (2006) 252–265.

22. X. Li, W. Mao, D. Zeng, F.-Y. Wang, Automatic construction of domain theory for attack planning, Proceedings of the 2010 IEEE International Conference on Intelligence and Security Informatics, 2010, pp. 65–70.

23. M. Newman, M. Girvan, Finding and evaluating community structure in networks, Physical Review E 69 (2; August) (2004) article 026113.

24. L. Tang, H. Liu, J. Zhang, Z. Nazeri, Community evolution in dynamic multi-mode networks, Proceedings of the Fourteenth ACM SIGKDD International Conference on Knowledge Discovery and Data Mining, 2008, pp. 677–685.

25. R. Albert, A. Barabasi, Statistical mechanics of complex networks, Reviews of Modern Physics 74 (1) (2002) 47–97.

26. F. Wu, B.A. Huberman, L.A. Adamic, J.R. Tyler, Information flow in social groups, Physica A: Statistical and Theoretical Physics 337 (1–2) (2004) 327–335.

27. G.B. Li, A. Lidia, S.V. Buldyrev, S. Havlin, H.E. Stanley, Transport and percolation theory in weighted networks, Physical Review E 75 (4) (2007) article 045103(R).

28. X. Li, W. Mao, D. Zeng, P. Su, F.-Y. Wang, Performance evaluation of machine learning methods in cultural modeling, Journal of Computer Science and Technology 24 (6) (2009) 1010–1017.

29. P. Su, W. Mao, D. Zeng, H. Zhao, Mining actionable behavior rules from group data, Proceedings of the IEEE 2011 International Conference on Intelligence and Security Informatics, 2011, pp. 181–183.

CHAPTER 8

Cultural Modeling for Behavior Analysis and Prediction

Cultural modeling is an emerging and promising research area in social computing [1–3]. How to investigate the impact of culture on human behavior or on social changes has become a popular research issue [4–8]. Cultural modeling aims to apply computational techniques to build behavioral models of organizations and analyze the impact of cultural factors on organizational behavior. It has raised interesting questions such as the discovery of the correlation between cultural factors and organizations' behavior, efficient and effective identification of the behavioral pattern of organizations from masses of cultural-related data, and the prediction of organizational behavior based on the cultural context.

In the meantime, significant opportunities exist for cultural research in security informatics given easily accessible content such as that from the Web. However, traditional methods rely heavily on human expert analysis, which is impractical for large amounts of data. New methodologies to efficiently analyze cultural dynamics from a wealth of security-related data are in great demand. The core issue in cultural modeling is the automatic construction of a group behavior model based on relevant cultural data. Machine learning methods play a critical role in such applications. The quality of the learned behavioral models as well as the accuracy of predictions rely heavily on the machine learning methods selected. As such, selection of a proper machine learning method is a crucial factor in cultural modeling.

The characteristics of cultural modeling datasets pose some major challenges from a machine learning perspective. Generally, cultural datasets are the historical behavioral record of organizations. They are often stored as a relational table for each organization. When a new organizational behavior or event occurs, the data items capture the current time, related cultural factors, and the state of the organization and behavior. Typically, the frequency of these events is low. In most cases, for a given organization, there are only of the order of dozens of records each year. Hence, the size of the datasets is always very small. At the same time, when certain behavior is observed, only a small number of attributes are identified. However, in order to analyze organizational behavior effectively, the number of attributes studied should be very large.

Advances in Intelligence and Security Informatics. 10.1016/B978-0-12-397200-2.00008-7

Based on the benchmark cultural dataset MAROB (Minorities at Risk Organizational Behavior) [9], in this chapter we investigate the performance of the major classification methods in cultural modeling, including naive Bayesian (NB), support vector machines (SVMs), artificial neural networks, k-nearest neighbor (kNN), decision trees, random forest (RF), and associative classification (AC). We evaluate these classification methods for their effectiveness in modeling cultural-related data and provide comparative analyses of algorithm performance in the security informatics domain.

8.1 Modeling Cultural Data in Security Informatics

Cultural and ideological conflicts have long been topics of great interest in intelligence and security informatics. The idea of cultural modeling was first proposed in Ref. [4]. Subrahmanian [4] suggested that real-time computational models can use various cultural factors to help policymakers predict the behavior of political, economic, and social groups. Subrahmanian et al. [5] then developed a conceptual cultural model called CARA (Cognitive Architecture for Reasoning about Adversaries). CARA is made up of four components: a Semantic Web extraction engine to elicit organization data; an opinion-mining engine that captures the group's opinions; an algorithm to correlate culturally relevant variables with the actions the organization takes; and a simulation environment within which analysts and users can experiment in hypothetical situations. When a new event occurs, CARA can forecast the k most probable sets of actions the group might take in the matter of a few minutes.

To further develop CARA, Khuller and colleagues proposed two algorithms, HOP and semiHOP, based on probabilistic logic programming and applied them in the terrorism domain [6]. By solving a set of linear programs, the algorithms are able to find the most probable actions (or sets of actions) of a certain organization in a possible world model. SemiHop can significantly reduce the number of variables in the linear programs and thus is faster than HOP. Nevertheless, there are still some weaknesses in these algorithms. For example, the improvement in computational efficiency is achieved at the cost of forecasting accuracy. Furthermore, these algorithms are not evaluated with any quantifiable measure, including accuracy.

To further improve efficiency and prediction quality, Martinez et al. [7] presented a new algorithm named CONVEX based on the MAROB datasets, which is more computationally efficient and accurate than HOP and semiHOP. CONVEX views each instance as a pair of two vectors, namely the context vector and action vector. The context vector contains the values of the context variables associated with the group, while the action vector contains the values of the action variables of the group. To predict the action vector, CONVEX uses distance functions in metric spaces to calculate the similarity between a given context vector and the other context vectors. Therefore, CONVEX is essentially a variant of the kNN method. The authors found from their experiments that when $k = 2$, using the Hamming or

Manhattan distance, CONVEX yields the best performance with an accuracy of 0.969. Although this result seems very good, it is not clear how well the same algorithm would perform in other meaningful measures. Furthermore, given the specific characteristics of cultural datasets, does the kNN-based method always yield the best performance in different measures? In this chapter, we look for answers to these questions via experimental studies.

8.2 Major Machine Learning Methods

8.2.1 Naive Bayesian (NB)

The naive Bayesian [10] is a classical probabilistic classifier based on Bayes' theorem. The NB classifier can be trained very efficiently in a supervised learning setting, depending on the precise nature of the probability model. It is viewed as an optimal classifier when there is no dependency between a particular feature and other features, given the class. Although simple, it has proved to be a successful classifier in a wide variety of domains.

8.2.2 Support Vector Machines (SVMs)

SVMs [11] are based on statistical learning theory and the Vapnik–Chervonenkis (VC) dimension. An SVM views the examples to be classified as two sets of vectors in an n-dimensional space and then builds a separating hyperplane in that space, maximizing the margin between the two datasets. In this study we use SMO, the first effective algorithm for SVMs, with the polynomial kernel.

8.2.3 Artificial Neural Networks

An artificial neural network [12,13] is composed of many simple processing elements (called artificial neurons) whose functionality is loosely based on the neurons in animal species. It learns via a process of adjustments to the connections between the processing elements and element parameters. As one of the most frequently used neural network architectures, multilayer perception (MLP) [13] is selected as a representative artificial neural network method for our experiment. We take the sum of the attribute count and the class count and divide it by 2 as the number of hidden layers.

8.2.4 k-Nearest Neighbor (kNN)

kNN is a simple but effective classification method. For a new instance to be classified, k nearest neighbors of the instances are selected, and then the major class of the k neighbors is assigned to the class of the new instance. When the kNN method is employed, the choice of k has a crucial effect on its classification performance. Martinez et al. [7] proposed a kNN-based algorithm (i.e. CONVEX) for the classification task. They found that CONVEX performed best on MAROB datasets when the k value was set to 2 and the Manhattan distance was chosen. In our experiment, we adopt the same parameter values to implement CONVEX as a representative of the kNN method.

8.2.5 Decision Trees

The most widely used classifier in machine learning is probably decision trees. Decision trees are tree structures that classify instances by sorting them based on feature values [14]. Within a decision tree, a node denotes the selected feature that is used to split input data and branches denote values of the node. Over the last few years, C4.5 has become a popular decision tree method. In our experiment, we chose C4.5 as a representative of decision trees.

8.2.6 Random Forest (RF)

Random forest [15] is a classifier that evolves from decision trees. It actually consists of many decision trees. To classify a new instance, each decision tree provides a classification for input data; random forest collects the classifications and chooses the most voted prediction as the result. The input of each tree is sampled data from the original dataset. In addition, a subset of features is randomly selected from the optional features to grow the tree at each node. Each tree is grown without pruning. Essentially, random forest enables a large number of weak or weakly-correlated classifiers to form a strong classifier.

8.2.7 Associative Classification (AC)

Associative classification [16] is a branch of data mining research that combines association rule mining with classification. Associative classification is a special case of association rule discovery in which only the class attribute is considered on the rule's right-hand side (consequent) [16]. After a large set of rules are generated, AC selects a subset of high-quality rules via rule pruning and ranking. Based on the reduced rule set, AC can then build an effective classifier. Associative classification is usually more accurate than the decision tree method in practice. The L3 associative classification algorithm [17] has been chosen for this experiment.

8.3 Experiment and Analysis

8.3.1 Datasets

In our experiment, we mainly use the MAROB cultural datasets for anti-terror research. The MAROB dataset is a subsidiary of the Minorities at Risk (MAR) Project [18], and covers

118 terrorism organizations in the Middle East ranging from 1980 to 2004. In the MAROB dataset, the historical data of each terrorist organization is represented as a relational database table, with the rows denoting years and the columns denoting behavioral attributes and culture attributes of the organization.

Behavioral attributes consist of violent behavior and nonviolent behavior. Violent behavior involves attack targets (e.g. attacks on civilians or government infrastructure) and different types of attack (e.g. kidnap, armed attack, etc.). Violent behavior comprises the class for prediction in our experiment. Nonviolent behavior involves different kinds of normal behavior, for instance negotiation with government, soliciting external support and so on.

Culture attributes include religious, economic, or political factors associated with an organization, such as religious belief and political grievances of the organization.

Although it comprises typical cultural modeling datasets, MAROB has some distinct characteristics. First of all, each set of organization data is rather sparse and the distribution of attribute values is imbalanced. Secondly, instances of each dataset are often small (with less than 100 instances per organization) and the number of attributes is extremely high (with more than 140 attributes per record). Our test is based on the 22 largest sets of organization data in the MAROB datasets.

8.3.2 Evaluation Measures

According to the characteristics of experimental data and domain features in cultural modeling, we chose three measures to evaluate the performance of the classifiers, namely accuracy, recall, and AUC, with 10-fold cross-validation.

Accuracy is commonly used in evaluating the quality of classification. However, there are some limitations to the accuracy measure in the cultural modeling domain. For example, accuracy cannot satisfactorily reflect a classifier's performance when the data distribution is imbalanced. In group behavior forecasting, we focus mainly on forecasting the occurrence of organizational behavior, hence the classification of positive examples is much more important than that of negative ones. In this case, overall high accuracy of an algorithm will not ensure its good performance in classifying positive examples.

Therefore, in addition to accuracy, we consider two other evaluation measures, recall and AUC. Recall is generally used for the data in which positive examples' misclassification costs are much higher than those of the negative ones. In our experiment, we focus on evaluating recalls of terrorism behavior that lead to destructive consequences.

AUC is the area under ROC curves. It is a statistically consistent and more discriminating measure than accuracy for balanced or imbalanced datasets [19]. We predict the organizational behavior with these three evaluation measures. For each algorithm, we report the average values of each measure and their standard deviations.

8.3.3 Experimental Results

Table 8.1 shows the average values and standard deviations of the accuracy, recall, and AUC measures for the seven experimental algorithms. The average values given in bold face correspond to the algorithm that has the highest average value, and the SD (standard deviation) in bold face corresponds to the algorithm that has the lowest standard deviation.

Tables 8.2 and 8.3 show the accuracy and AUC values of the prediction for different attribute sets. For each algorithm, the lowest accuracy and AUC values are marked in bold.

8.3.4 Observations and Analysis

It can be seen from Table 8.1 that all the algorithms achieve high accuracies of more than 90%, whereas the recalls are rather low and AUCs vary considerably from classifier to classifier. This is partly due to the skewed distribution of the experimental data. Most of the datasets have a large ratio of negative examples, in contrast to the small percentage of positive examples, which results in high accuracies of performance yet a lower level of recalls. The higher the degree of class imbalance problem and the smaller the

Table 8.1 Overall Experimental Results of the Seven Algorithms

Measure		Algorithm						
		NB	SMO	MLP	CONVEX	C4.5	RF	L3
Accuracy	Ave.	**0.969**	0.941	0.952	0.944	0.929	0.940	0.929
	SD	**0.044**	0.060	0.053	0.052	0.069	0.056	0.063
AUC	Ave.	**0.978**	0.697	0.915	0.854	0.530	0.788	0.866
	SD	**0.054**	0.216	0.168	0.249	0.401	0.227	0.154
Recall	Ave.	**0.778**	0.420	0.525	0.344	0.362	0.428	0.430
	SD	**0.303**	0.434	0.409	0.405	0.404	0.404	0.431

Table 8.2 Accuracies of Prediction for Different Attribute Sets

Attribute	Accuracy (%)						
	NB	SMO	MLP	CONVEX	C4.5	RF	L3
Attack types	0.965	**0.932**	**0.941**	**0.929**	**0.922**	0.927	0.921
Transnational violence outside the country	0.981	0.947	0.953	0.950	0.926	**0.922**	**0.911**
Transnational violence within the country	0.993	0.968	0.979	0.969	0.947	0.965	0.956
Domestic violence	0.969	0.956	0.952	0.958	0.936	0.946	0.942

Table 8.3 AUCs of Prediction for Different Attribute Sets

Attribute	AUC (%)						
	NB	SMO	MLP	CONVEX	C4.5	RF	L3
Attack types	**0.964**	0.662	**0.890**	0.839	0.351	0.757	**0.840**
Transnational violence outside the country	0.995	0.616	0.945	**0.806**	0.354	0.696	0.887
Transnational violence within the country	0.995	**0.578**	0.915	0.859	**0.340**	**0.651**	0.908
Domestic violence	0.977	0.681	0.925	0.881	0.464	0.746	0.872

overall size of the training set, the more classifiers will be sensitive to the class imbalance problem [20].

The naive Bayesian significantly outperforms the other algorithms in terms of accuracy, AUC, and recall, with the lowest standard deviations. Although the attributes in our datasets do not satisfy the independence assumption, there are two main reasons for naive Bayesian showing the best performance. As pointed out above, the MAROB datasets are very small (only 30–60 instances for an organization). The naive Bayesian often performs very well on smaller datasets [21]. The other reason is that zero-one loss is used as loss function in our experiment, and the naive Bayesian is optimal under zero-one loss (misclassification rate) even when the independence assumption is violated by a wide margin [21].

From Table 8.1, the C4.5 algorithm is almost always dominated by other algorithms. This is not surprising because decision trees are prone to errors in classification problems with many classes and a relatively small number of training examples [22]. As a variant of the decision tree algorithm, the random forest algorithm gives better recall and much better AUC results than C4.5, indicating that the performance of ensemble classifiers may be superior to a single classifier. The associative classifier outperforms C4.5 with a recall of 0.068 and an AUC of 0.336. As the AUC measure represents the overall performance of a classifier, this is expected because the associative classifier has proved to be much better than C4.5 [17,23–25].

Note that kNN (the CONVEX algorithm used in the related work is actually a variant of kNN) performs the worst in terms of average recall. One explanation for this is that kNN is dominant in relatively dense datasets [26], but for highly spare datasets its performance suffers as it is unable to form reliable neighborhoods. Our experimental data are indeed sparse and the number of positive examples is much less than that of negative ones. The positive examples are so sparse that it is very difficult for kNN to find positive neighbors for positive examples. As a result, kNN's recall in this case is very low. Furthermore, kNN seems to be more sensitive to the class imbalance problem. Research also shows that SVMs and MLP are not as sensitive to imbalance as kNN [20,27–29].

As shown in Tables 8.2 and 8.3, the accuracy and AUC performances of the seven classifiers in different attribute sets reveal interesting results. Five of the seven classifiers give the lowest accuracy in the 'attack types' attribute set. In addition, three of the seven classifiers give the lowest AUC performance in the same attribute set. In our experiment, the attributes used for prediction are grouped into four types. 'Attack types' means the attack modes of the terrorists, for instance bombing, hijacking, and kidnap. 'Transnational violence outside the country' means that the organization is using violence to target transnational entities outside the country of the organization as a strategy. 'Transnational violence within the country' means that the organization is using violence to target transnational entities within the country of the organization as a strategy. 'Domestic violence' means the organization is using violence domestically as a strategy. The bad prediction performance of attack types indicates that the attack types of the terrorists are hard to predict depending on the cultural attributes defined in the experimental data. Alternatively, we can use the occurrence of attack, the severity of terrorism attack, and some other attributes to replace attack types.

8.4 Discussions on Cultural Modeling Research Issues

Our experimental studies show that the prediction results can be further improved in modeling cultural data for security-related applications. Below, we further discuss cultural modeling research issues in constructing datasets, selecting features and learning algorithms, model interpretability, and incorporating domain knowledge.

8.4.1 Cultural Datasets Construction

The first step, and a key difficulty in cultural modeling, is the construction of structural datasets from an open source. This difficulty arises from two factors. First, manually collecting relevant raw data from the large amount of information on the Web is rather painstaking. This needs the participation of a number of people and considerable effort. Second, the selection of appropriate attributes relating to an organization is quite problematic. Furthermore, some attributes in cultural datasets are difficult to quantify, for example the friendship between two countries. These tasks require substantial domain knowledge, which usually relies on the efforts of domain experts. Due to these difficulties, the agent-based approach may be a promising alternative to represent and model social cultural situations.

8.4.2 Attribute Selection

We notice that some cultural factors remain constant in an organization dataset, for example religious belief always remains the same for a particular organization. In this case, these attributes do not contribute to the classification. These cultural factors are inherent properties of an organization. They should comprise the cultural context of the organization rather than being used for classification. Along this line of thinking, each organization should have a cultural context, which is composed of various cultural attributes. Cultural context is part of the cultural level of an organization and is beyond its historical record. We can further learn an organization's behavioral pattern from certain cultural factors by comparing different organizations' behavioral patterns and cultural context.

8.4.3 Best Performance of Classifiers

Given the fundamental data characteristics in cultural modeling, the performance of classifiers varies according to other specific features of the datasets. Consider the optimality condition of the naive Bayesian as an example. It has been proved that NB is optimal for any two-class concept with nominal features under zero-one loss [30–32]. As all the environmental attributes in the MAROB datasets are nominal and all the forecasting behavioral attributes are two-class concepts, it is not surprising that NB performs best in these experiments. However, for the more complex cultural datasets, where the optimality condition of NB is no longer satisfied, NB may not be the best classifier in all the measures.

8.4.4 Handling the Class Imbalance Problem

Experimental results show that in the cultural modeling domain the class imbalance problem has a significant impact on the performances of the learning algorithms in the cultural modeling domain. Research [20,28,33] has proposed several methods to tackle this problem, such as oversampling, undersampling, and cost modifying. Oversampling and undersampling are mainly based on the sampling of original datasets. These two methods are more effective for large datasets. The cost-modifying method consists of modifying the relative cost associated with misclassifying the positive and the negative classes so that it compensates for the imbalance ratio of the two classes [20]. Since our datasets are rather small, we plan to adopt the cost-modifying method to handle the class imbalance problem in our future work.

8.4.5 Model Interpretability

An important issue in cultural modeling research is how to interpret the resulting models. We need to know the association between cultural factors and organizational behavior in an explainable manner. Classifiers should not only derive statistical models for behavior prediction, but also provide a reasonable explanation of the behavior. Therefore, rule-based classifiers (e.g. C4.5 and the associative classifier in our study) are preferable to cultural modeling. C4.5 can induce decision rules for expert evaluation. The associative classifier can generate more association rules, and its comprehensibility is better than C4.5. Despite their mediocre performances, these two classifiers are superior to other classifiers in the interpretability of models.

8.4.6 Incorporation of Domain Knowledge

As the results show, the performances of the evaluated learning methods are still far from ideal. Recalls of all the algorithms are fairly low. This shows that pure data-driven approaches may not be sufficient in predicting organizational behavior. We can explicitly utilize domain knowledge to improve the prediction results in cultural modeling. Related work in data mining also suggests similar improvements [34,35]. The incorporation of domain knowledge will also improve the interpretability of the prediction models.

8.4.7 Cultural and Social Dynamics of Behavioral Patterns

The cultural-related data we adopt is sorted by year. It is possible that during the time span, the cultural and social backgrounds underlying the data have changed greatly. As a result, behavioral patterns extracted from the historical data may not correctly reflect current and future trends of group behavior. How to effectively track the evolution of behavioral patterns has become an important research issue [36].

8.5 Conclusion

Cultural modeling has gained increasing attention in recent social computing and security informatics research. In this chapter, we compare the performance of seven machine learning methods for behavior prediction and analysis. We test these classifiers on cultural datasets, and use three measures to evaluate and analyze the experimental results. The results show that the naive Bayesian method performs best across all configurations. However, all the algorithms that we investigated (including the variant of kNN used in the related work) have relatively low recalls and AUCs varying from classifier to classifier. This suggests that pure data-driven classification methods may not fully meet the performance needs in behavior modeling and prediction in intelligence and security informatics. The findings motivate us to incorporate domain knowledge, agent modeling, and a probabilistic reasoning approach into group behavior modeling, analysis, and prediction.

References

1. F.-Y. Wang, K.M. Carley, D. Zeng, W. Mao, Social computing: From social informatics to social intelligence, IEEE Intelligent Systems 22 (2) (2007) 79–83.
2. F.-Y. Wang, Toward a paradigm shift in social computing: The ACP approach, IEEE Intelligent Systems 22 (5) (2007) 65–67.

3. D. Zeng, F.-Y. Wang, K.M. Carley, Social computing, IEEE Intelligent Systems 22 (5) (2007) 20–22.

4. V.S. Subrahmanian, Computer science: Cultural modeling in real time, Science 317 (5844) (2007) 1509–1510.

5. V.S. Subrahmanian, M. Albanese, M.V. Martinez, D. Nau, D. Reforgiato, G.I. Simari, A. Sliva, J. Wilkenfeld, O. Udrea, CARA: A cultural-reasoning architecture, IEEE Intelligent Systems, 22 (2) (2007) 12–16.

6. S. Khuller, V. Martinez, D. Nau, G. Simari, A. Sliva, V.S. Subrahmanian, Finding most probable worlds of logic programs, Proceedings of the First International Conference on Scalable Uncertainty Management, 2007, pp. 45–59.

7. V. Martinez, G.I. Simari, A. Sliva, V.S. Subrahmanian, CONVEX: Context vectors as a paradigm for learning group behaviors based on similarity, IEEE Intelligent Systems 23 (4) (2007) 51–57.

8. F.-Y. Wang, Is culture computable? IEEE Intelligent Systems 24 (2) (2009) 2–3.

9. Minorities at Risk Organizational Behavior Dataset, Minorities at Risk Project, Center for International Development and Conflict Management, University of Maryland, College Park, 2008, <http://www.cidcm.umd.edu/mar>.

10. D.J. Hand, K. Yu, Idiot's Bayes—Not so stupid after all? International Statistical Review 69 (3) (2001) 385–398.

11. V.N. Vapnik, Support vector estimation of functions (Part II), Statistical Learning Theory, John Wiley & Sons Inc., (1998) 375–570.

12. A.K. Jain, J. Mao, K.M. Mohiuddin, Artificial neural networks: A tutorial, Computer 29 (3) (1996) 31–44.

13. C.M. Bishop, Neural Networks for Pattern Recognition, Oxford University Press, 1995.

14. S. Kotsiantis, I. Zaharakis, P. Pintelas, Machine learning: A review of classification and combining techniques, Artificial Intelligence Review 26 (3) (2006) 159–190.

15. L. Breiman, Random forests, Machine Learning 45 (1) (2001) 5–32.

16. F. Thabtah, A review of associative classification mining, Knowledge Engineering Review 22 (1) (2007) 37–65.

17. E. Baralis, P. Garza, A lazy approach to pruning classification rules, Proceedings of the Second IEEE International Conference on Data Mining, 2002, pp. 35–42.

18. Minorities at Risk Project, Center for International Development and Conflict Management, College Park, Maryland, 2005, <http://www.cidcm.umd.edu/mar>.

19. C.X. Ling, J. Huang, H. Zhang, AUC: A statistically consistent and more discriminating measure than accuracy, Proceedings of the Eighteenth International Joint Conference on Artificial Intelligence, 2003, pp. 329–341.

20. N. Japkowicz, S. Stephen, The class imbalance problem: A systematic study, Intelligent Data Analysis 6 (5) (2002) 429–450.

21. R. Kohavi, D.H. Wolpert, Bias plus variance decomposition for zero-one loss functions, Proceedings of the Thirteenth International Conference on Machine Learning, 1996, pp. 275–283.

22. M. Govindarajan, Text mining technique for data mining application, Proceedings of World Academy of Science, Engineering and Technology 26 (104) (2007) 544–549.

23. B. Liu, W. Hsu, Y. Ma, Integrating classification and association rule mining, Proceedings of the Third International Conference on Knowledge Discovery and Data Mining, 1998, pp. 80–86.

24. W. Li, J. Han, J. Pei, CMAR: Accurate and efficient classification based on multiple class-association rules, Proceedings of the First IEEE International Conference on Data Mining, 2001, pp. 369–376.

25. X. Yin, J. Han, CPAR: Classification based on predictive association rules, Proceedings of the Third SIAM International Conference on Data Mining, 2003, pp. 369–376.

26. M. Grčar, D. Mladenič, B. Fortuna, M. Grobelnik, Data sparsity issues in the collaborative filtering framework, Proceedings of the Seventh International Workshop on Knowledge Discovery on the Web, 2005, pp. 58–76.

27. J. Nathalie, Class imbalances: Are we focusing on the right issue? Proceedings of the ICML'2003 Workshop on Learning from Imbalanced Data Sets, 2003.

28. N. Japkowicz, The class imbalance problem: Significance and strategies, Proceedings of the Second International Conference on Artificial Intelligence, 2000, pp. 111–117.

29. J. Zhang, kNN approach to unbalanced data distributions: A case study involving information extraction, Proceedings of the ICML'2003 Workshop on Learning from Imbalanced Data Sets, 2003.

30. I. Rish, J. Hellerstein, J. Thathachar, An analysis of data characteristics that affect naive Bayes performance, Technical Report, IBM T.J. Watson Research Center, 2001.

31. P. Domingos, M. Pazzani, On the optimality of the simple Bayesian classifier under zero-one loss, Machine Learning 29 (2–3) (1997) 103–130.

32. M. Martinez-Arroyo, L. E. Sucar, Learning an optimal naive Bayes classifier, Proceedings of the Eighteenth International Conference on Pattern Recognition, 2006.

33. R. Barandela, J.S. Sanchez, V. Garcia, E. Rangel, Strategies for learning in class imbalance problems, Pattern Recognition 36 (3) (2003) 849–851.

34. C. Zhang, P.S. Yu, D. Bell, Domain-driven data mining, IEEE Transactions on Knowledge and Data Engineering 22 (6) (2010) 753–754.

35. L. Cao, C. Zhang, Domain-driven, actionable knowledge discovery, IEEE Intelligent Systems 22 (4) (2007) 78–88.

36. D. Nau, J. Wilkenfeld, Computational cultural dynamics, IEEE Intelligent Systems 23 (4) (2008) 18–19.

Index

Page numbers followed by "f" indicate figures, and "t" indicate tables.

Academic Press is an imprint of Elsevier
The Boulevard, Langford Lane, Kidlington, Oxford OX5 1GB, UK
225 Wyman Street, Waltham, MA 02451, USA

First edition 2012

Notice
No responsibility is assumed by the publisher for any injury and/or damage to persons or property
as a matter of products liability, negligence or otherwise, or from any use or operation of any
methods, products, instructions or ideas contained in the material herein. Because of rapid
advances in the medical sciences, in particular, independent verification of diagnoses and drug
dosages should be made.

British Library Cataloguing in Publication Data
A catalogue record for this book is available from the British Library

Library of Congress Cataloging-in-Publication Data
A catalog record for this book is available from the Library of Congress

ISBN–13: 978-0-12-37200-2

Printed and bound in the US

12 13 14 15 16   10 9 8 7 6 5 4 3 2 1

Printed in the United States
by Bookmasters

Printed in the United States
By Bookmasters